2009 台达杯国际太阳能建筑设计竞赛获奖作品集
Awarded Works from International Solar Building Design Competition 2009

阳光与希望
Sunshine and Hope

中国可再生能源学会太阳能建筑专业委员会　编
Edited by Special Committee of Solar Building, CRES

执行主编　仲继寿　张磊
Chief Editor: Zhong Jishou, Zhang Lei

翻译　王岩
Translator : Wang Yan

中国建筑工业出版社
China Architecture & Building Press

图书在版编目(CIP)数据

阳光与希望／中国可再生能源学会太阳能建筑专业委员会编．—北京：中国建筑工业出版社，2009
2009台达杯国际太阳能建筑设计竞赛获奖作品集
ISBN 978-7-112-10928-9

Ⅰ.阳… Ⅱ.中… Ⅲ.太阳能住宅-建筑设计-作品集-中国-现代 Ⅳ.TU241.91-64

中国版本图书馆 CIP 数据核字(2009)第058595号

本作品集由"台达环境与教育基金会"赞助出版

责任编辑：唐　旭
版式设计：付金红
责任校对：兰曼利　梁珊珊

2009 台达杯国际太阳能建筑设计竞赛获奖作品集
Awarded Works from International Solar Building Design Competition 2009
阳光与希望
Sunshine and Hope
中国可再生能源学会太阳能建筑专业委员会　编
Edited by Special Committee of Solar Building, CRES
执行主编　仲继寿　张磊
Chief Editor: Zhong Jishou, Zhang Lei
翻译　王岩
Translator: Wang Yan

*

中国建筑工业出版社出版、发行(北京西郊百万庄)
各地新华书店、建筑书店经销
北京图文天地制版印刷有限公司制版
北京中科印刷有限公司印刷

*

开本：787×1092毫米　1/12　印张：28　字数：640千字
2009年5月第一版　2009年5月第一次印刷
印数：1—2500册　定价：98.00元
ISBN 978-7-112-10928-9
　　　　(18166)

版权所有　翻印必究
如有印装质量问题，可寄本社退换
(邮政编码 100037)

谨将本书献给5·12中国汶川大地震灾区的孩子们和建设者

感谢台达环境与教育基金会资助举办2009国际太阳能建筑设计竞赛

感谢台达电子集团资助阳光小学建设，暨重建中国四川省绵阳市涪城区杨家镇小学

感谢所有关心阳光小学建设事业的人们——

This book is dedicated to the children and builders of 5·12 Earthquake Area in Wenchuan, China.

Thanks to Delta Environmental & Educational Foundation that aids financially International Solar Building Design Competition 2009.

Thanks to Delta Electronics, Inc. that aids financially the construction of sunshine primary school and reconstruction of Yang Jia Zhen School, Fucheng District, Mianyang, Sichuan Province, China.

Thanks to all the people concerned about the construction of sunshine primary school.

目 录
CONTENTS

阳光与希望　Sunshine and Hope — 008

过程回顾　General Background — 009

2009台达杯国际太阳能建筑设计竞赛评审专家介绍
Introduction of Jury Members of International Solar Building Design Competition 2009 — 014

一、综合奖作品　General Prize Awarded Works — 001

一等奖　First Prize

蜀光　Light of Shu — 002

土生土长　Earth & Growth — 008

二等奖　Second Prize

暖暖　NuanNuan — 014

种子　Seed — 018

阳光小学　Solar Elementary School — 024

时光游廊　The Sunshine Corridor of Time — 030

三等奖　Third Prize

绵阳阳光小学规划设计　ICNRUS — 036

长屋　Long House — 042

爱 & 阳光 & 希望　Love & Sunshine & Hope — 048

普适的地域性　Universal Regionality — 054

舞动的藏袍　Flowing Tibet Robe — 060

气候与塑形　Forms Follows Climate — 064

优秀奖　Honorable Mention Prize

光之墨迹　Sunshine, Ink Marks	068
绵阳地区阳光小学竞赛设计　Mianyang Solar Primary School Design	072
阳光与希望　Sunshine and Hope	076
阳光 & 温暖　Light & Warm	080
绿野　Green Field	086
553号作品　No.553	090
绵阳阳光小学　The Sunshine Primary School in Mianyang	096
光织品　Light Fabric	100
阳光小学　Sunshine Primary School	104
童年的梦想　The Dream of the Childhood	108
应变建筑　Climate-responsive building	112
进化的小学　Evolution	118
光阴趣事　Sunshine Shadow & Fun	122
光·盒　Sunshine & Square	126
时光 阳光 童年　Time Sunshine Childhood	130
"L"代表什么？　What is "L"？	136
希望 绿色 生态　Hope Green Ecologica	142
阳光"家"希望　Sun+Hope=Home	148
日光播放器　Solar Player	154
缓冲盒　The Buffer Box	160
生态学校　Eco-School	166
马尔康地区阳光小学　Rural Sunshine Primary School In Ma Er Kang Area	170

马尔康之花　Blossom at Ma Er Kang	176
370号作品　No.370	180
一半是记忆，一半是未来　Half is memory, half is future	184
阳光 & 梦想　Sunshine & Dream	188
阳光面对面　Face To Sunshine	192
阳光温室　Sunlight Greenhouse	196
阳光猎人　Sunshine Hunter	200
阳光小学设计　Sunshine Primary School Design	204
534号作品　No.534	208
阳光下的格桑花　Sunshine Gesang Flower	214
太阳-斜面　Solar-Slope	218
分享阳光　Share Sunshine	222
阳光·乐园　Sunny Garden School	226
阳光街小学　The Solar Road	230
阳光容器　A Container Of Sunshine	234
感光　Sense The Sunlight	238
薪·火相传　Firewood Devolution	244
巢　Nest	250

二、技术专项奖作品　Prize for Technical Excellence Works

阳光与希望　Sunshine & Hope	256
太阳伞　Solar Umbrella	262
绵阳阳光小学　The Sunshine Primary School in Mianyang	266

家·园·院	My Family & My School & My Paradise	270
集合标准建筑	Assembled-Standard Construction	276
农村阳光小学	Rural Sunshine Primary School	282
呼吸	Breathe	288

参赛人员名单　　Name List of Attending Competitors　　292

2009台达杯国际太阳能建筑设计竞赛办法
Competition Brief for International Solar Building Design Competition 2009　　301

绵阳市涪城区杨家镇台达阳光小学设计方案说明
Explanation about the Design Scheme of Delta Sunshine Primary School in Yang Jia Zhen, Fucheng District, Mianyang　　312

后记　　Postscript　　317

阳光与希望
Sunshine and Hope

太阳高高挂，

孩子快快长。

太阳带来温暖，

孩子带来快乐。

太阳照遍世界，

孩子走遍未来。

没有太阳和孩子就没有世界，

没有太阳和孩子就没有未来。

我们就是太阳和小孩！

这是一个四年级的小学生创作的诗歌，也诠释了2009台达杯国际太阳能建筑设计竞赛的主题——阳光与希望。

本次竞赛将目光投向5•12汶川地震灾区，面向全球征集农村"阳光小学"设计方案。"阳光小学"不仅强调绿色与环保理念，而且注重建筑技术与太阳能利用技术的融合。竞赛活动试图通过以太阳能被动利用为主、主动利用为辅的建筑设计，为农村地区的孩子们营造安全健康、可持续运营的校园。

考虑到地震给灾区学校造成的巨大损失，竞赛组委会决定按照灾区典型气候特征（中国四川马尔康地区和绵阳地区）制定设计条件，吸引全球更多的设计团队参与竞赛活动，用才智打造"阳光小学"，用行动传播太阳能建筑理念，将温暖洒向灾区孩子的心灵。

本次竞赛吸引了来自四大洲33个国家和地区的共1024个团队注册，征集有效作品194份。评审专家们对获奖作品给予了充分的肯定，认为作品普遍强调了建筑技术与太阳能技术的有机结合，在经济可行、技术可靠的前提下，具有一定的超前性。

国际太阳能建筑设计竞赛活动已经被国际太阳能学会批准成为两年一届的常规赛事，并配合世界太阳能大会定期举行。展望2011，我们会一如既往地努力搭建好竞赛舞台，让更多优秀的设计师对太阳能等可再生能源和建筑的结合进行更深入的尝试和探索，让更多优秀的作品通过大家的努力落地开花，让未来的建筑拥有更广阔的遐想空间。

感谢2009台达杯国际太阳能建筑设计竞赛活动的参与者，感谢所有关心与支持太阳能建筑发展的人们。

科技引领时代，创意点亮未来。让我们一同传递爱心与责任，播种梦想与希望。

Sun is hanging in the sky,
Children are growing fast.
Sun brings warmth,
Children bring happiness.
Sun lightens the whole world,
Children go to the future.
World disappears without sun and children,
Future disappears without sun and children.
Hi! We are the sun and children!

This is a poetry by a four grade pupil, which also explains the theme of the International Solar Building Design Competition 2009 – Sunshine and Hope.

In this Competition we put our sight to the disaster area of 5·12 Earthquake happened in Wenchuan and collect design schemes of rural "Sunshine Primary School" world widely. For the "Sunshine Primary School", not only the idea of green and environmental protection is emphasized but also more attention is paid to good combination of building technology and solar application technology. The competition tries to build a safe, healthy and sustainable operating school for the children in rural area by means of building design with passive solar application mainly and active as auxiliary.

Considering the huge loss of the schools by the earthquake, the Organizing Committee decided to make design conditions according to the typical climate character of the disaster area (Ma Er Kang area and Mianyang area of Sichuan province) and attracted more design teams of the world to attend competition activity which may create a "Sunshine Primary School" by their intelligence, diffuse the idea of solar buildings by their action and besprinkle the warmth to the heart of the children.

In this competition, 1024 teams and groups from 33 countries and areas in the world are attracted to take registration and 194 effective works are collected. The jury members have given full affirmation to awarded works and deemed that the works emphasize organic combination of building technology and solar technology and they have certain advancing if operable in economy and liable in technology.

The activity of the international solar building design competition has been approved by International Solar Energy Society as a conventional match of two-year-session, which will be taken place regularly coordinated with Solar World Congress. In prospect of 2011, we will do our best to build a better match stage as in the past giving opportunity to more outstanding designers to make deeper trying and exploration on the combination of renewable energies with buildings. It is expected that more excellent works will achieve in implementation and architecture and have more spaces of fancy in future.

Thanks to all participants of the International Solar Building Design Competition 2009 and all the people concerned about and supporting the development of solar buildings.

The science and technology lead the age and the originality lightens the future. Let's transfer our love and responsibly and seed the dream and hope.

过程回顾
General Background

学校重建场地　（The site for reconstruction of the School）

本次竞赛由国际太阳能学会和中国可再生能源学会联合主办；国家住宅与居住环境工程技术研究中心、中国可再生能源学会太阳能建筑专业委员会承办；台达环境与教育基金会独家冠名。在各单位的通力协作下，竞赛组委会于2008年1月成立，并组织了竞赛启动、媒体宣传、校园巡讲、建校选址、作品注册与提交、作品初评与终评、技术交流会等一系列活动。这些活动得到了海内外业界人士的积极响应和参与，参赛作者更是用他们辛勤的劳动将近200件多姿多彩的答卷呈现在我们面前。

一、竞赛筹备

自筹备之初，竞赛组委会就将本届竞赛策划为获奖作品可以实地建设，以此为目标把较易实施的农村学校锁定为竞赛题目，在山东、江西、甘肃等地开展了选址活动。2008年5月12日中国汶川发生震惊世界的大地震，灾区孩子们的教学环境引起各界的普遍关注。竞赛组委会决定结合灾后重建工作，为那里的孩子们建造一所太阳能等清洁能源支持的阳光小学，让他们在安全、舒适、环保、节能的环境下健康成长。

在台达环境与教育基金会、四川省绵阳市台湾事务办公室的大力斡旋下，组委会组织专家到绵阳市进行了实地考察，并最终将杨家镇小学确定为设计竞赛的场地建设条件，编制了竞赛设计任务书。

二、竞赛启动

2008年6月25日，2009台达杯国际太阳能建筑设计竞赛在京启动。本次竞赛结合中国汶川地震的灾后重建工作，以"阳光与希望"为主题，向全球征集农村"阳光

The competition is organized by International Solar Energy Society (ISES) and Chinese Renewable Energy Society (CRES) and operated by China National Engineering Research Center for Human Settlements (CNERCHS) and Special Committee of Solar Buildings, CRES. Delta Environmental and Educational Foundation is the sponsor. With a concerted effort of all sponsors, the Organizing Committee of the competition was established in January, 2008 and had organized a series of activities concerning the competition, such as start-up, media publicity, lecture tour, selection of school site, registration and submission of the works, preliminary evaluation and final evaluation, technological intercourse, etc. All activities gained active response and participation from the professionals at home and abroad. Moreover, through the hard efforts, nearly 200 authors presented their brilliant and colorful examining papers.

1. Preparation

In the beginning the organizing Committee had set up a target, which was to take awarded works into construction and made sure the project would be a rural school that might be easy to be implemented. And then the activities of site selection were started around some provinces, such as Shandong, Jiangxi, Gansu, etc. On May 12, 2008 a world-shaking earthquake happened in Wenchuan of Sichuan Province while the teaching environment problem for the children in disaster area had arisen serious attention of people. The Organizing Committee decided to build a sunshine school for them combined with the re-construction after the earthquake, which would be supported by clean energies such as solar energy and make the children grow in a safe, comfortable, well-protected and energy-saving environment.

With the great endeavor from Delta Environmental & Educational Foundation and Taiwan Affair Office of Mianyang, Sichuan Province, the organizing committee organized the relevant experts to review the site in Mianyang, and finally Yang

发布会出席嘉宾留影（VIP photo in the start-up conference）

小学"设计方案，竞赛题目分为马尔康地区农村阳光小学和绵阳地区农村阳光小学两项，参赛人员可任选一项进行设计。赛后，部分获奖方案将在灾区等地付诸建设。

三、校园巡讲

2008年9月至12月，竞赛组委会组织进行了为期3个月的"2009台达杯国际太阳能建筑设计竞赛"校园巡讲活动，参加的高校包括重庆大学（共52组参赛队伍）、清华大学（共30组参赛队伍）、天津大学（共10组参赛队伍）、浙江大学（共8组参赛队伍）和华南理工大学（共5组参赛队伍）。巡讲的内容包括太阳能建筑应用现状、发展前景及往届竞赛获奖作品介绍等，均受到了在校师生的热烈欢迎。注册人数在巡讲后均有明显上升。

Jiazhen primary school was chosen to be the construction location.

2. Start-up

International Solar Building Design Competition 2009 was started up on June 25, 2008 in Beijing. Combined with the re-construction after the earthquake of Wenchuan in China with a theme of "Sunshine and hope" the design schemes concerning rural "sunshine school" were started to collect design schemes from the whole world. The subjects of the competition were rural "Sunshine primary school" in Ma Er Kang area and Mianyang area, one of which could be chosen for the participants. As the competition finished part of awarded schemes would be taken into construction in disaster area.

重庆大学现场（2008年10月15日）
Lecture tour in Chongqing University (Oct.15, 2008)

浙江大学现场（2008年12月11日）
Lecture tour in Zhejiang University (Dec.11, 2008)

清华大学现场（2008年11月14日）
Lecture tour in Tsinghua University (Nov.14, 2008)

华南理工大学现场（2008年12月12日）
Lecture tour in South China University of Technology (Dec.12, 2008)

天津大学现场（2008年11月27日）
Lecture tour in Tianjin University (Nov.27, 2008)

四、媒体宣传

在专业技术巡讲的同时，自2008年6月至2009年1月组委会开展了多渠道的媒体宣传工作，包括：2008年6月正式开通竞赛双语网站实时报告竞赛进展情况并开展太阳能建筑的科普宣传，在www.google.com刊登关键字搜索广告，以便社会大众更快捷地登陆竞赛网站；在中国《建筑学报》，日本《新建筑(Shinkenchiku)》、《A+U》，韩国 SPACE magazine，美国《AR》等近10家国内外专业杂志；新华网、雅虎网、ABBS、日本AKICHIATLAS.com、德国Architects24.com、美国ArchNewsNow.com、意大利architecture. It、丹麦bygnet.dk等国内外75家网站；以及《科技日报》、《中国建设报》、《环球时报》等30余家国内的平面媒体，向国际发布了竞赛的组织与宣传情况。

3. Lecture tour in campuses

From September to December, 2008 the Organizing Committee organized an activity of lecture tour concerning the competition in many colleges and universities for three months. The participants were from Chongqing University (52 teams), Tsinghua University (30 teams), Tianjin University (10 teams), Zhejiang University (8 teams) and South China University of Technology (5 teams). The lectures are including existing situation about solar energy application in buildings, the developing prospect and introduction about awarded works of previous competitions. They were warmly welcomed by teachers and students in campuses. After lecture tour the registration numbers of the participants of the competition were increasing evidently.

国外杂志
Foreign magazines

五、援建协议签署仪式

2008年12月4日，2009台达杯国际太阳能建筑设计竞赛选址进展新闻发布会在北京召开。竞赛组委会在会上宣布，将运用台达电子集团于汶川地震后捐赠的1000万元人民币，援助四川省绵阳市涪城区的杨家镇小学，进行全校异地重建。根据重建计划，杨家镇"阳光小学"施工图，将于2009年5月，由中国建筑设计研究院等建筑设计单位，根据建设要求完善本次竞赛获奖作品而成，并于2009年6月开始付诸施工建设，最终将于2010年春季开学前建成并启用。

六、竞赛注册情况汇总

本次竞赛的注册时间为2008年6月25日至12月18日，共1024个团队注册，其中：国内（包括港、澳、台地区）866个团队，国际158个团队，涉及的国家和地区有：中国、美国、加拿大、墨西哥、厄瓜多尔、阿根廷、意大利、葡萄牙、法国、俄罗斯、德国、英国、瑞典、西班牙、荷兰、丹麦、希腊、乌克兰、捷克、波兰、芬兰、澳大利亚、新西兰、伊朗、印度、日本、巴基斯坦、马来西亚、韩国、印度尼西亚。

4. Media publicity

At the same time with the professional lecture tour, from June, 2008 to January, 2009 the Organizing Committee developed the work of media publicity in multi-channel. In June, 2008 a web site concerning the competition in English and Chinese was set up to put forward the real-time report of competition process and do some propaganda about science popularization. Through web site of Google a search advertisement of key word was published, so that people might visit competition web quickly. Besides, the situation of organization and propaganda of the competition was promulgated to the world via near 10 professional journals as Architecture Journal of China, "Shinkenchiku" of Japan, A+U, SPACE magazine of Korea, AR, etc, 75 web sites as Xinhuanet, Yahoo, ABBS, AKICHIATLAS.com, Architects24.com, ArchNewsNow.com, architecture. It, bygnet.dk and more than 30 domestic print medias as Science and Technology Daily, China Construction News, Globe Times, etc.

5. Ceremony of agreement subscription regarding the aid project

On Dec. 4, 2008 a press conference about the site selection of the school was taken place in Beijing. The Organizing Committee announced that RMB 10 millions contributed after the earthquake by Delta Electronic Group would be used for the

七、作品提交情况汇总

截止到2009年1月18日，大赛组委会共收到国内外所提交的作品204份，包括有效作品194份，无效作品10份，其中涉及绵阳地区作品108份，马尔康地区作品86份。作品提交情况如下：

中国内地：168
    院校：139（83%）
    设计院、设计公司及个人：29（17%）
港澳台地区：9
    院校：9（100%）
    设计院、设计公司及个人0（0%）。
国外：27
    院校：7（26%）
    设计院、设计公司及个人：20（74%）。

涉及的国家有：
美洲国家（占22%）包括：美国，加拿大，墨西哥；欧洲国家（占56%）包括：意大利，葡萄牙，法国，俄罗斯，德国，瑞典，西班牙，捷克，波兰；亚洲国家（除中国外）（占22%）包括：日本，伊朗。

八、作品初评

2009年1月20日，大赛组委会将全部有效作品提交给初评专家组，经过评审专家近20天紧张而高效的初评工作，每位专家根据竞赛办法中规定的评比标准对每一件作品进行评审，各自选出60份进入终评的作品。组委会对所有专家的评审结果进行统计后，整理出综合得票数最多的前60份作品进入终评阶段。

九、作品终评

2009年台达杯国际太阳能建筑设计竞赛终评会于2009年2月23～24日在北京中国建筑设计研究院召开。经专家组讨论，一致推选德国不伦瑞克理工大学建筑与太阳能技术学院院长M.Norbert Fisch教授担任本次终评工作的评审组长。在他主持下，专家组活动按照简单多数的原则集体讨论和公正客观地评选作品，最终选出2项一等奖作品（绵阳及马尔康地区各一件）、4项二等奖作品、6项三等奖作品以及优秀奖作品40项。

reconstruction of Yang Jia Zhen School in Fucheng District, Mianyang, Sichuan province which would be rebuilt in a new place. According to the reconstruction plan, the construction drawings of Yang Jia Zhen "sunshine school" will be completed in May, 2009 by China Architecture Design & Research Group and others based on the awarded works revised and perfected according to the demands of the construction. The project will be started in construction in June, 2009 and finished and taken into use before term beginning in the spring, 2010.

6. Registration

The registration time of the competition is from June 25 to December 18, 2008. There are 1024 groups or teams registered. They are 866 of China (including Hong Kong, Macau and Taiwan) and 158 from abroad dealing with following countries and areas: China, USA, Canada, Mexico, Ecuador, Argentina, Italy, Portugal, France, Russia, Germany, UK, Sweden, Spain, the Netherlands, Greece, Ukraine, Czech, Poland, Finland, Australia, New Zealand, Iran, India, Japan, Pakistan, Malaysia, Korea and Indonesia.

7. Work submission

Up to January 18, 2009 the Organizing Committee has received 204 works from the whole world including 194 of effective ones and 10 of non-effective ones and 108 works designed for Mianyang area while 86 for Ma Er Kang area.

It can be subdivided as follows.

168 works come from China mainland:
    139 works (83%) from colleges and universities
    29 works (17%) from institutes or companies of building design and individuals.

9 works from Hong Kong, Macau and Taiwan:
    9 works (100%) from colleges and universities
    0 work (0%) from institutes or companies of building design and individuals.

27 works come from abroad:
    7 works (26%) from colleges and universities
    20 works (74%) from institutes or companies of building design and individuals.

Overseas works come from America (22%) including USA, Canada and Mexico, Europe (56%) including Italy, Portugal, France, Russia, Germany, Sweden, Spain, Czech and Poland and Asia (22%) including Japan and Iran.

8. Preliminary evaluation

On January 20, 2009 all effective works were submitted to the jury of preliminary evaluation by the Organizing Committee. The work of preliminary evaluation was going on tensely and effectively for near 20 days. Every jury member gave appraisal to every work according to the evaluation standard of the competition and chose 60 works for final evaluation. And then the Organizing Committee made a statistics of the evaluation result of the jury and arranged an order of the works. Top 60 of works with the most votes would enter the final evaluation.

终评会场一角 (Glance in the final evaluation conference)　　现场讨论 (Discussion in the confernce)

十、组织国际可再生能源建筑集成技术交流会

为配合竞赛终评会，大赛组委会于2009年2月25日上午，在中国建筑设计研究院举办"国际可再生能源建筑集成技术交流会"，由德国评委M.Norbert Fisch教授介绍了德国在非住宅建筑节能方面的示范工程和经验，来自台湾的绿色建筑专家林宪德教授，则详细介绍了由台达电子集团董事长郑崇华先生个人捐建的台南成功大学"绿色魔法学校"的太阳能创新设计及绿色建筑材料的特色。与会者包括国内建筑师、规划师、工程师等与相关建筑技术研究人员及部分参赛团队。

2009年6月结合绵阳市涪城区杨家镇小学重建工程的开工仪式，竞赛组委会将在绵阳召开隆重的竞赛颁奖典礼。表彰获奖团队的同时，也期待能有更多的获奖作品可以用于灾后重建当中，让阳光为那里的孩子们带来希望。

9. Final evaluation

Final evaluation of the international of solar building design competition 2009 was taken place in February 23~24, 2009 in Beijing at China Architecture Design & Research Group. By discussion of the jury Mr. M. Norbert Fisch, Professor of TU Braunschweig, President of the Institute of Architecture and Solar Energy Technology, Germany was elected as the director of the jury. Presided by him all jury members discussed and evaluated the works fairly and impersonally. Finally the awarded works were selected out including two of First Prizes (one in Mianyang and other one in Ma Er Kang area), four of Second Prizes, six of Third Prizes and 40 of Honorable Mentioned Prizes.

10. International Technical Exchange Seminar of Building Integrated Renewable Energy

Coordinated with the conference of final evaluation of the competition, a technical exchange seminar of building integrated renewable energy was taken place on February 25, 2009 by the Organizing Committee in China Architecture Design & Research Group. Prof. M.Nobert Fisch, jury member from Germany, had an introduction about demonstrated projects and experience of energy saving for German non-housing buildings. Prof. Lin Xinde, expert of green building of Taiwan, introduced the solar innovative design in Tai Nan Success School, "Green Magic School", and the characteristics of green materials as well, which is distributed Mr. Bruce Cheng, board chairman of Delta Electronic Group. Domestic architects, town planners, engineers, relevant research personnel of building technology and part of team members of the competition attended the seminar.

In June, 2009 the Organizing Committee will hold a ceremony of awarding prizes of the competition in the site of Yang Jia Zhen school project of Fucheng District, Mianyang together with the ceremony of construction starting. It is expected that more of awarded works can be applied in reconstruction of the disaster area and the sun shine will bring more hope to the children there.

美术教师和绘画的孩子们　　鸟瞰震后杨家镇小学

2009台达杯国际太阳能建筑设计竞赛评审专家介绍
Introduction of Jury Members of International Solar Building Design Competition 2009

评审专家：
JURY MEMBERS:

崔恺：国际建筑师协会副理事，中国建筑学会副理事长，中国国家工程设计大师，中国建筑设计研究院总建筑师。

Mr.Cui Kai, Deputy Board Member of IUA (International Union of Achitects); Vice President kf Architectural Society of China; National Design Master and Chief Architect of China Architecture Design & Research Group.

Anne Grete Hestnes女士：前国际太阳能学会主席，挪威科技大学建筑系教授。

Ms.Anne Grete Hestnes, Former President of International Solar Energy Society and Professor of Department of Architecture, Norway Science & Technology University.

Deo Prasad Pacific：国际太阳能学会亚太区主席，澳大利亚新南威尔士大学建筑环境系教授。

Mr.Deo Prasad, Asia-Pacific President of International Solar Energy Society (ISES) and Professor of Faculty of the Built Environment, University of New South Wales, Sydney, Australia.

M.Norbert Fisch：德国不伦瑞克理工大学教授(Tu Braunschweig)，建筑与太阳能技术学院院长，德国斯图加特大学博士。

Mr.M.Norbert Fisch, Professor of TU Braunschweig, President of the Institute of Architecture and Solar Energy Technology, Germany and Doctor of Stuttgart University, Germany.

Mitsuhiro Udagawa：国际太阳能学会日本区主席，日本早稻田大学博士，日本工学院大学建筑系教授。

Mr.Mitsuhiro Udagawa,President of ISES-Japan;Doctor of Engineering of Waseda University and professor of Department of Architecture,Kogakuin University.

林宪德：台湾绿色建筑委员会主席，日本东京大学博士，台湾成功大学建筑系教授。

Mr.Lin Xiande,President of Taiwan Green Building Committee;Doctor of Tokyo University,Japan and Professor of Faculty of Architecture of Success University,Taiwan.

仲继寿：中国可再生能源学会太阳能建筑专业委员会主任委员，中国矿业大学博士。

Mr.Zhong Jishou,Chief Commissioner of Special Committee of Solar Building, Chinese Renewable Energy Society and Doctor of China University of Mining & Technology.

喜文华：甘肃自然能源研究所所长，联合国工业发展组织国际太阳能技术促进转让中心主任，联合国可再生能源国际专家，国际协调员。

Mr.Xi Wenhua,Director-General of Gansu Natural Energy Research Institute;Director-General of sustainable energy field from United Nations,international coordinator.

冯雅：中国建筑西南设计研究院副总工程师，中国建筑学会建筑热工与节能专业委员会副主任，重庆大学博士。

Mr.Feng Ya,deputy chief engineer of Southwest Architecture Design and Research institute of China;deputy director of special committee of building thermal and energy efficiency,Architectural Society of China,Doctor of Chongqing University.

一、综合奖作品
General Prize Awarded Works

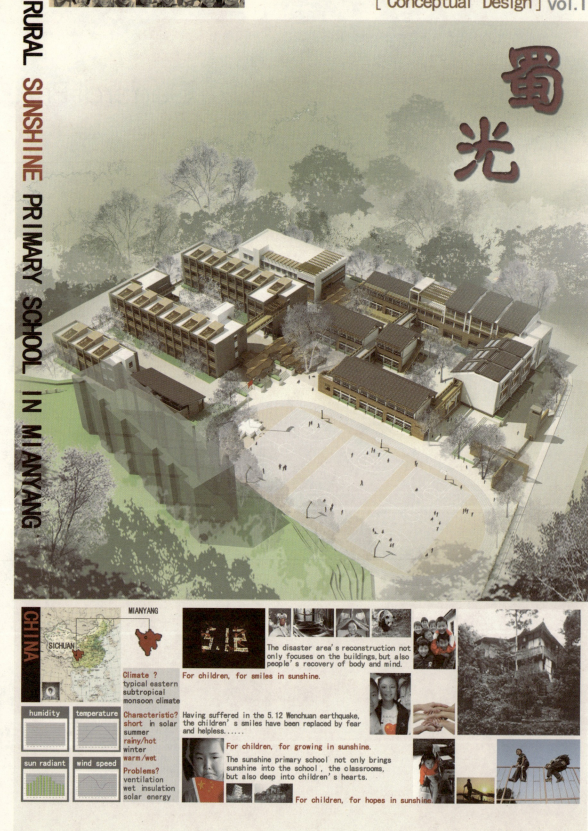

一等奖
First Prize

项目名称：蜀光
        Light of Shu
作　　者：刘慧、孙瑞、贾培斌、李献良、
        张增武
参赛单位：山东建筑大学建筑城规学院

专家点评：
住宿区、教学区和运动区布置合理；开放的空间和围合的庭院景观为室外活动增添了乐趣；从学校入口到大门再到庭院，形成一个简洁的流动空间；不同分区间的联系紧密而又不互相干扰；设计具有地方文化特色，恰当地运用了当地的材料，如竹廊架、木屋架等；食堂布置在教学区与住宿区之间比较合理。
宿舍北向设置走廊缓冲空间，房屋朝向良好；屋檐部分采用半透明的镂空手法，利于室内采光；在两片蓄热的墙体间设置了空气层，利于保温隔热；双层屋面结构的设计防止了夏季室内过热；采用了雨水收集系统，节约用水。
建议在食堂和宿舍的屋面采用主动太阳能技术，来提供沐浴的热水。

Good zones for living, teaching and sports for children;Very interesting open spaces and court garden for outdoor activities; Good connection and circulation between different spaces;Good local culture design and local material application with bamboo frame and wood structure;Good canteen arrangement between teaching and living areas;Single bank with a north aisle arrangement from dormitory rooms with good orientation for passive solar gain.
Good daylight design with semi-transparent overhang roofs;Using air layer between the two walls and heating the air;Using double roof construction to prevent overheating in summer;Using rain water collection technology;It can be improved by using active solar heaters on the roofs of canteen and dormitory for bath.

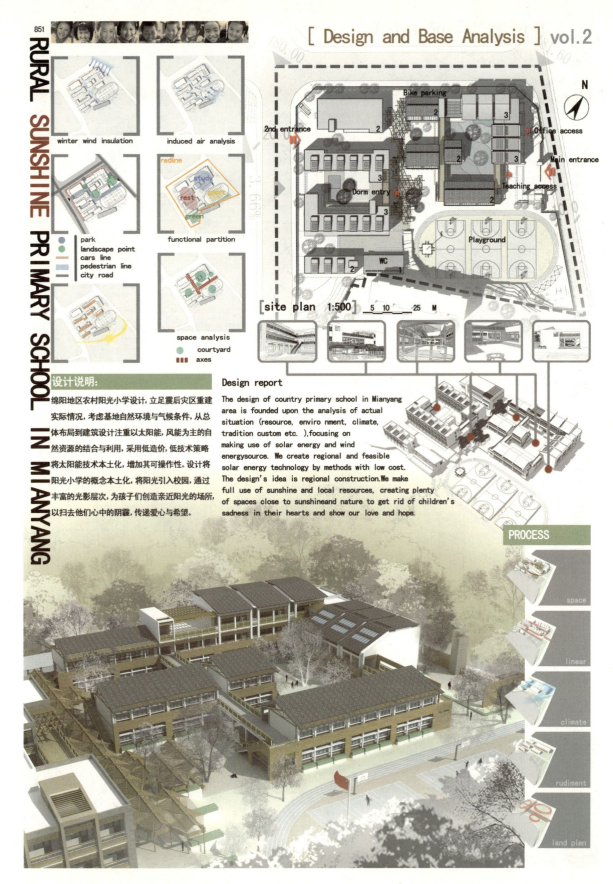

RURAL SUNSHINE PRIMARY SCHOOL IN MIANYANG

[Design and Base Analysis] vol.2

2009 台达杯国际太阳能建筑设计竞赛获奖作品集

- winter wind insulation
- induced air analysis
- functional partition

space analysis
- park
- landscape point
- cars line
- pedestrian line
- city road
- courtyard
- axes

[site plan 1:500]

设计说明：

绵阳地区农村阳光小学设计，立足震后灾区重建实际情况，考虑基地自然环境与气候条件，从总体布局到建筑设计注重以太阳能，风能为主的自然资源的结合与利用，采用低造价，低技术策略将太阳能技术本土化，增加其可操作性。设计将阳光小学的概念本土化，将阳光引入校园，通过丰富的光影层次，为孩子们创造亲近阳光的场所，以扫去他们心中的阴霾，传递爱心与希望。

Design report

The design of country primary school in Mianyang area is founded upon the analysis of actual situation (resource, environment, climate, tradition custom etc.), focusing on making use of solar energy and wind energysource. We create regional and feasible solar energy technology by methods with low cost. The design's idea is regional construction. We make full use of sunshine and local resources, creating plenty of spaces close to sunshine and nature to get rid of children's sadness in their hearts and show our love and hope.

PROCESS
- space
- linear
- climate
- rudiment
- land plan

003

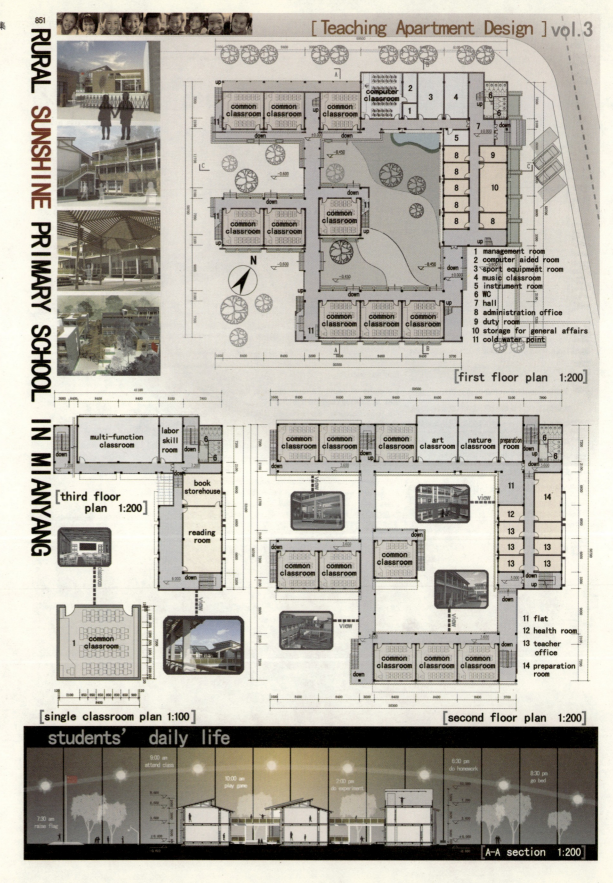

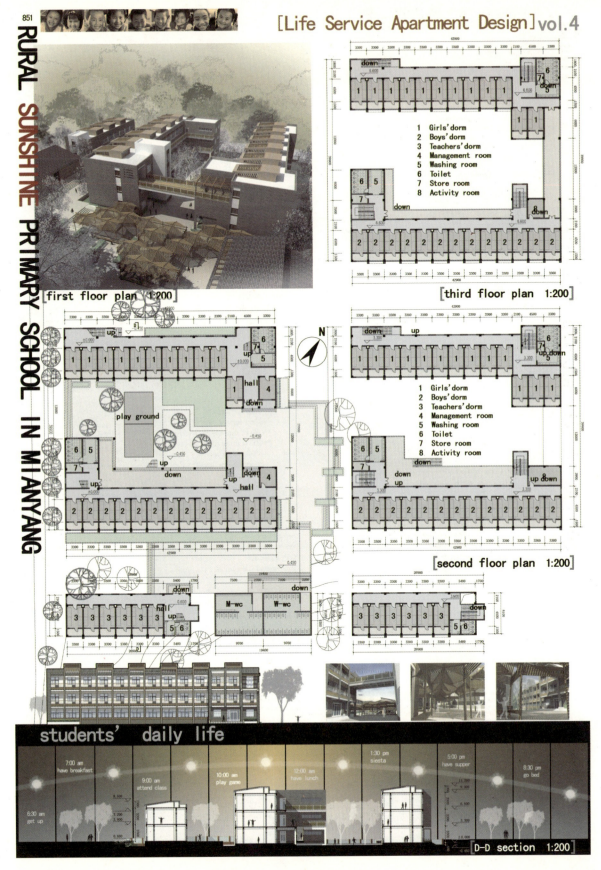

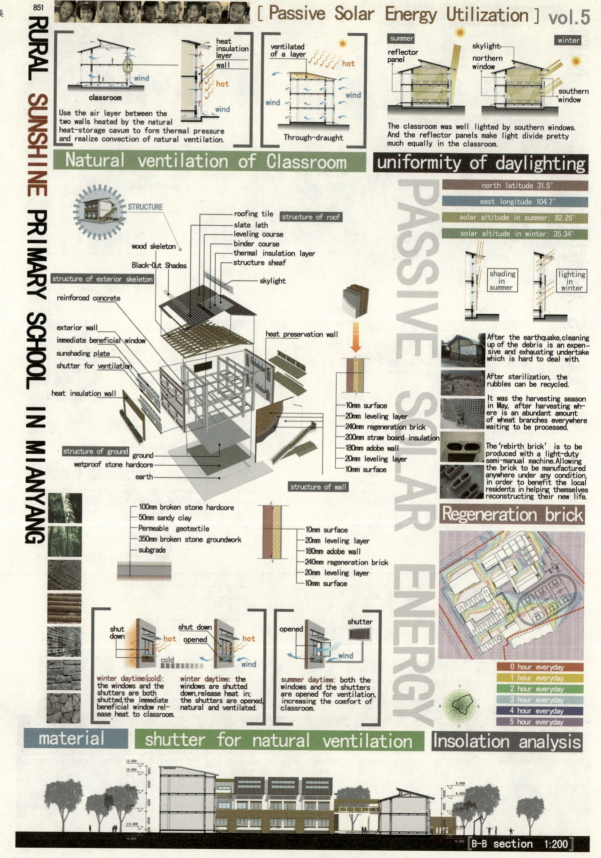

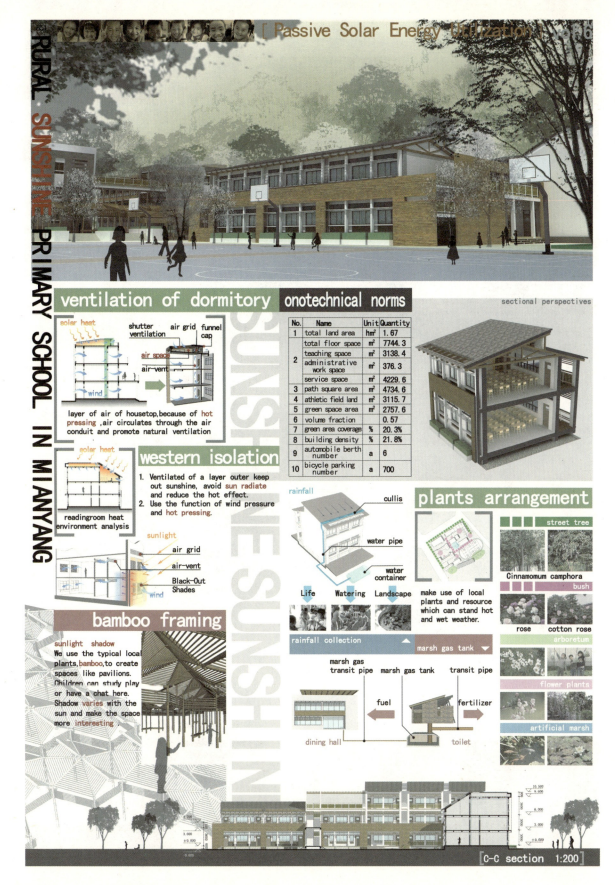

一等奖
First Prize

项目名称：土生土长
　　　　　Earth & Growth

作　　者：李晓东、刘文、侯韵、孙燕怡、
　　　　　李媛媛、鞠晓磊、李浩田

参赛单位：山东建筑大学建筑城规学院

专家点评：

住宿区、教学区和运动区布置合理；设计思路与马尔康地区寒冷的气候紧密结合；挡风坡地的设计，使得建筑能够躲避冬季冷风；采用南向的直接受益窗、厚重的结构墙体、北向走廊缓冲空间等被动式手段；建筑形体的设计能够在冬季得到良好的太阳辐射量，而在夏季又不使室内变得过热；采用太阳能烟囱来加强自然通风；采用太阳能集热器进行地板辐射采暖。

Good arrangement for teaching, living and sports areas;Compact arrangement for the cold climate;Good underground earth space design to make shelter form the cold winter wind;Good design in open south side window, massive interior walls and the north corridor for passive solar gains;Good building shape design to get solar radiation in the winter and reduce overheating in the summer;Using wind tower design to support the natural ventilation; Using solar thermal collector for floor heating.

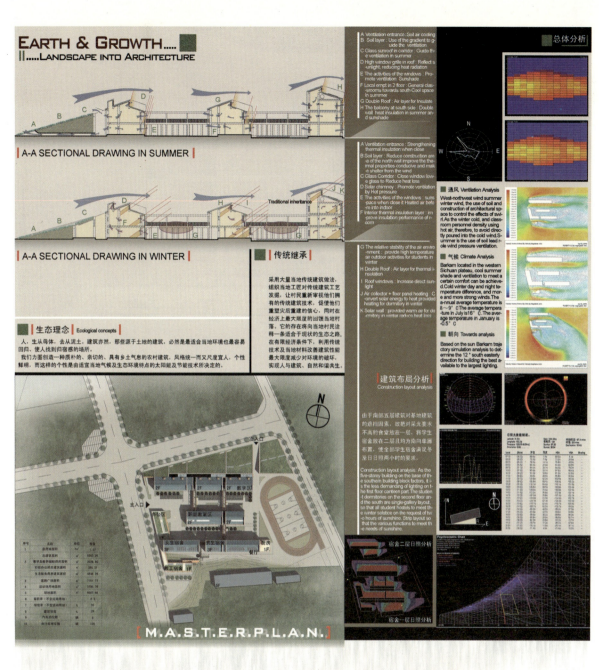

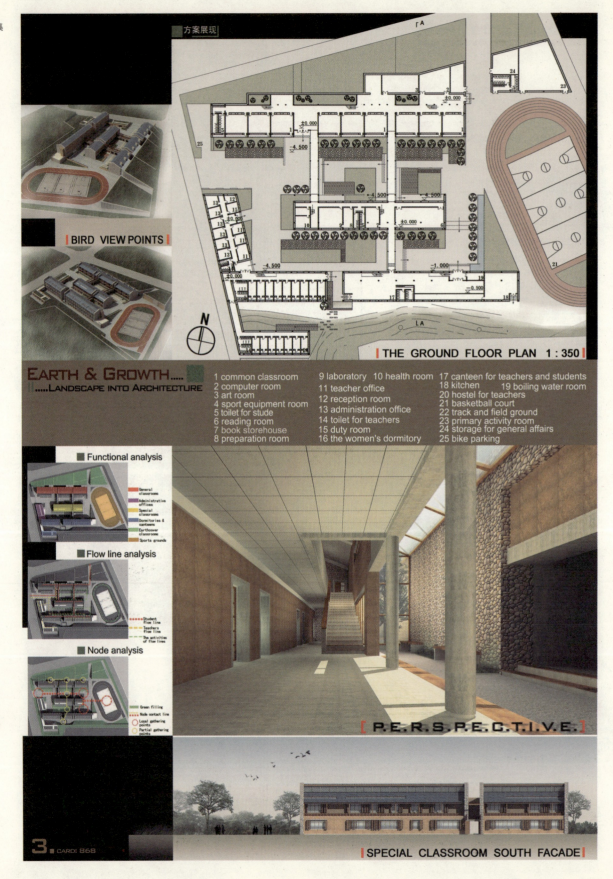

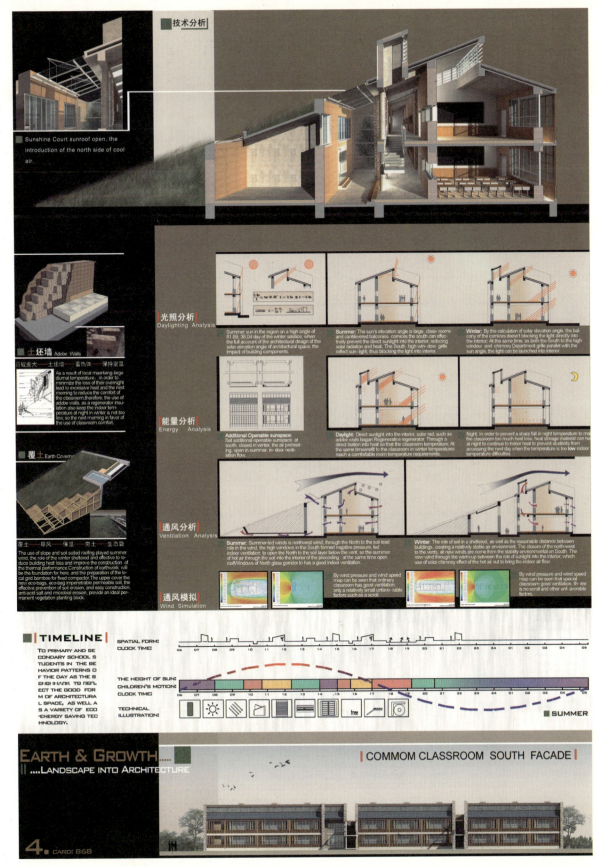

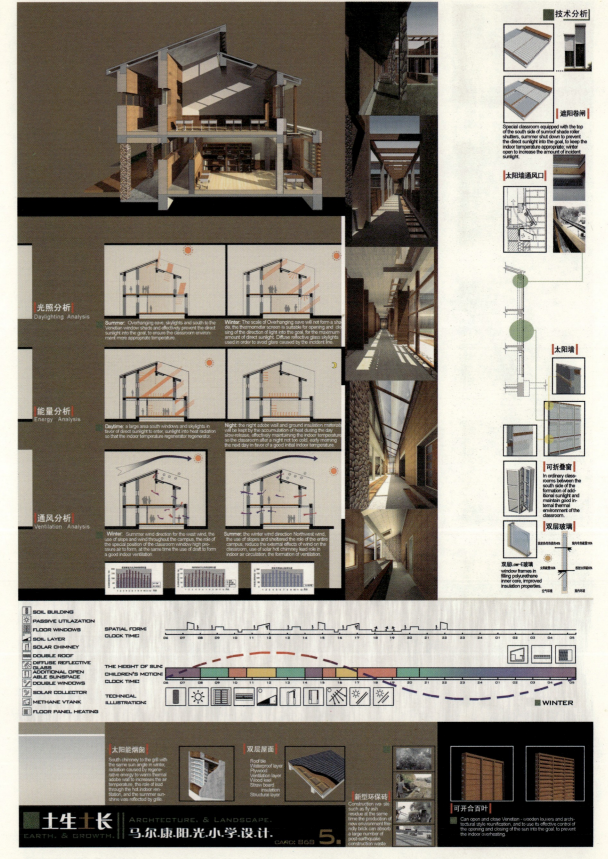

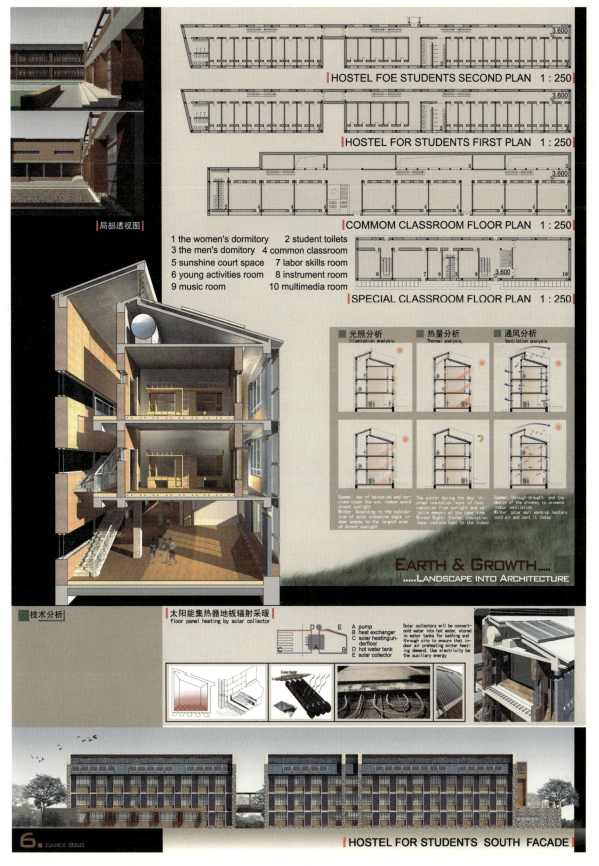

二等奖
Second Prize

项目名称：暖暖
　　　　　NuanNuan
作　　者：顾芳、刘碧峤
参赛单位：华中科技大学建筑与城市规划学院

专家点评：

功能分区合理，建筑风格也比较有当地特色；除了利用太阳能外，还考虑了沼气等其他可再生能源的利用；雨水收集系统的采用节约了用水；大量的背景分析和计算给设计提供了很充分的理论依据。

The functional zoning is reasonable and the architectural style has local character comparatively. Except solar energy the utilization of other renewable energies such as biogas is also considered in the scheme. Rain water collection system is adopted, which is advantaged to save water. A big amount of background analysis and calculation provides adequate theoretical basis.

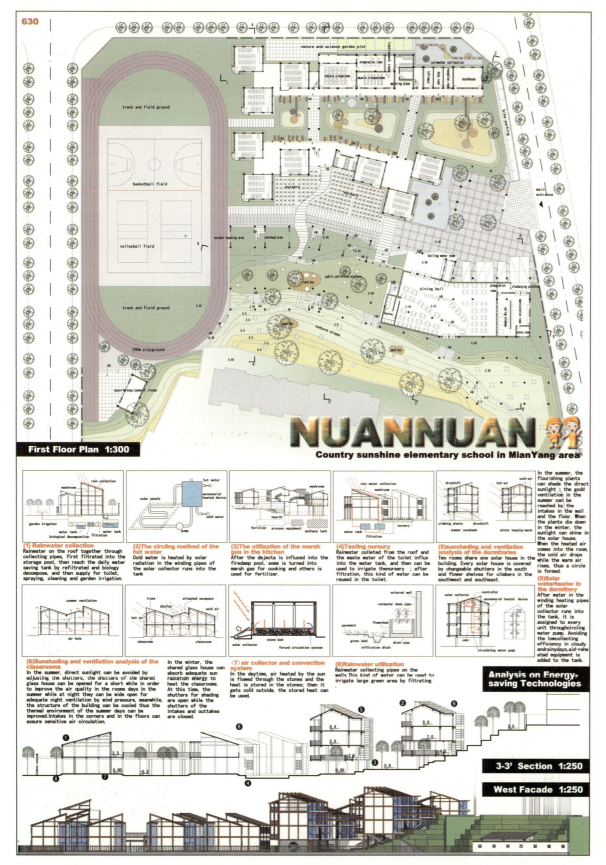

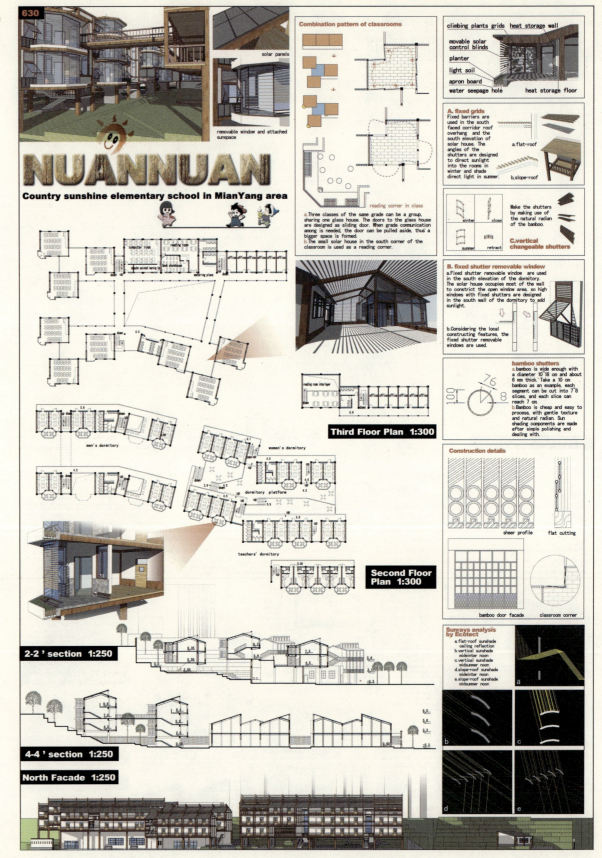

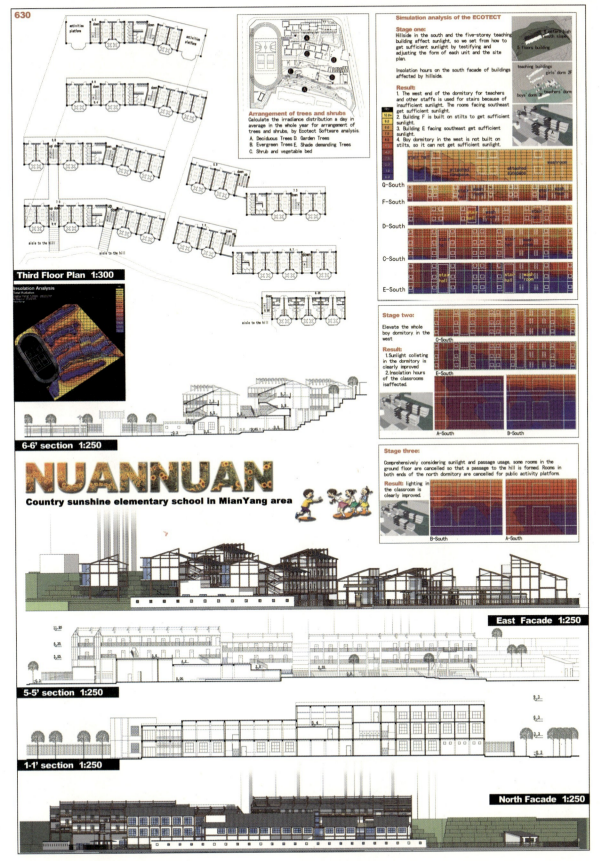

2009台达杯国际太阳能建筑设计竞赛获奖作品集　　注册编号 0857 绵阳地区小学设计

二等奖
Second Prize

项目名称：种子
　　　　　Seed
作　者：孙铭、郑恒祥、杨磊、宋安、
　　　　刘筱、杨琳
参赛单位：山东建筑大学建筑城规学院

专家点评：

该方案在操场等各个方面的设计很有想法；总体布局很合理；很多想法都通过各种手段表达了出来；利用太阳能技术很成功；从整体上看，冬、夏分区很直观，冬天阳光斜射，夏天尽量不进入；建筑形态和设计与当地文化结合得很好；在利用材料方面，用到了当地的材料、双层玻璃，还有地震废弃物，就地取材；绿地比较多，除了被动式还采用了一些主动式的技术。

It is full of ideas in all aspects including playground in this scheme. The master layout is very reasonable. A lot of ideas have a good expression by all kinds of measures. It is successful on utilization of solar technology. In the whole, the adaptability to the winter and the summer is visual directly and easily that in the winter the sunlight will shine in and in the summer the sunlight will be kept to a minimum to entering the rooms. The architectural form and design are good in combination with local culture. In aspect of material utilization local materials, double glass windows as well as castoff in earthquake are utilized. More of green field are arranged. Except passive solar energy some technology concerning active solar energy is adopted as well.

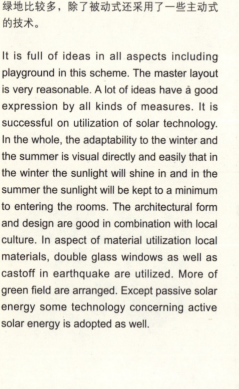

设计说明　DESIGN NOTES

方案建筑单体设计中以"种子"为概念，体现了方案的节能性，生态性以及生长性。这些集保温，通风等多种太阳能节能技术于一体的"种子"散落在整个校园，而"种子"作为教学区的生态角，办公区的休息厅，宿舍区的室外活动晒台等，其功能更多样化，不仅改善了建筑小气候而且还增加了建筑的趣味性。其"种子"采取可变模式，其夏季开敞利于通风，冬季闭合利于保暖。

方案建筑设计，建筑入口广场，中心庭院，室外剧场区明确，流线互不干扰，空间相互渗透，宿舍底层架空，使教学区和生活区之间的庭院和广场小气候大大改善，并为学生提供多种半室内，半室外的活动空间，也使宿舍避免了阴坡的影响，使其室内日照远远高于标准，生活区与教学区通过连廊相连，使其空间连续。考虑绿色建筑设计，庭院自然园地广场，绿化等以当地气候的变化而设计。考虑可持续建筑设计方案提供一个可适应各地的万能布局，并为建筑再生提供可塑的条件。

The conception of energy saving, eco-friendly and the easy possibility of further construction in this project are interpreted by the seed-like construction units. These so-called seeds are sowed everywhere on the campus with the functions of thermal insulation, ventilation and so on in the light of solar energy utilization technology. These units can play the roles of eco-corner, resting rooms of office area and open air playgrounds in dorm zones. Despite their diverse functions, the seed units modify the microclimate, and at the same time, make the construction more interesting. All seeds are variable, they can be opened for ventilation in summers and shut tight for thermal insulation in winters.

In the project, the division of the entrance square, the central yard and the outdoor theater are quite clear, stream lines don't interfere with each other, spaces are naturally connected, the first floor of the dorm building rise off the ground, the space between them greatly modifies the microclimate in the yards and squares amid teaching and living areas and provide the students with more semi-open playgrounds. This , at the same time,offsets the negative influence of the southern slope, filling the rooms with more than enough sunlight. The corridors connect the spaces between the teaching and living areas. The natural park and planting are designed according to local climate change. Such sustainable construction planning set an example of universal arrangement, and makes the reproduction of architectures a possibility.

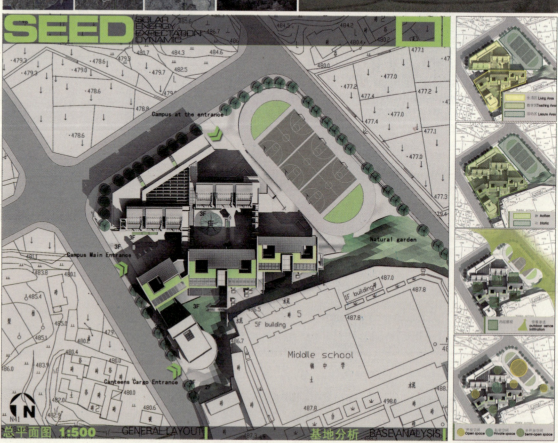

018

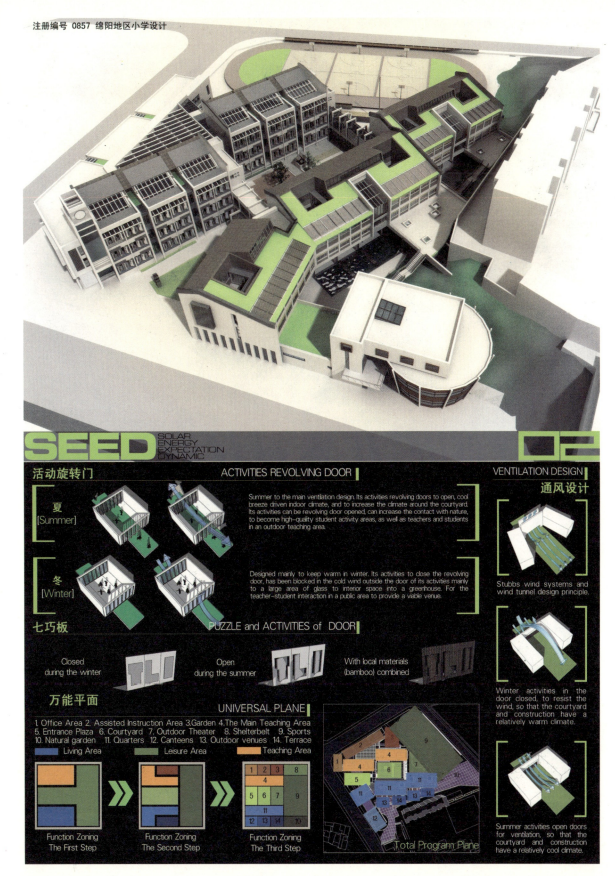

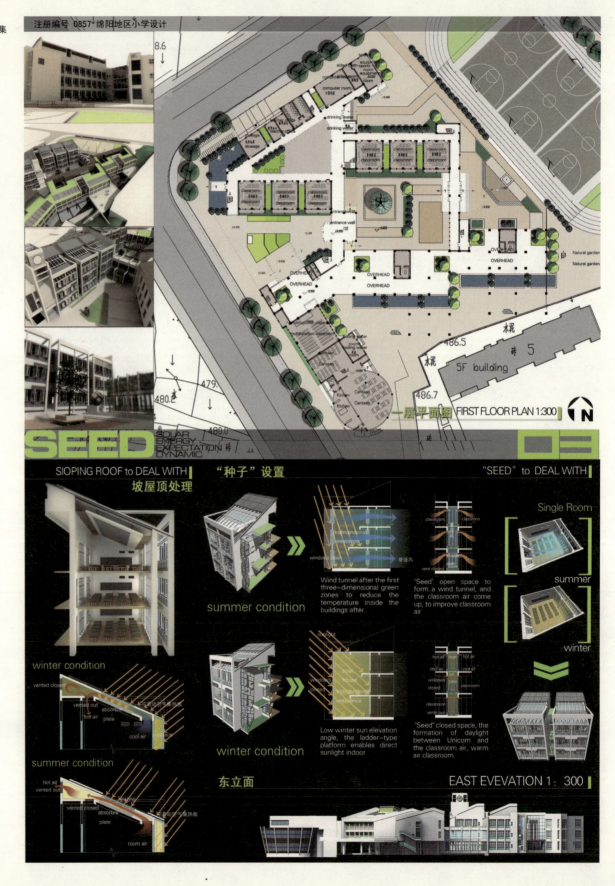

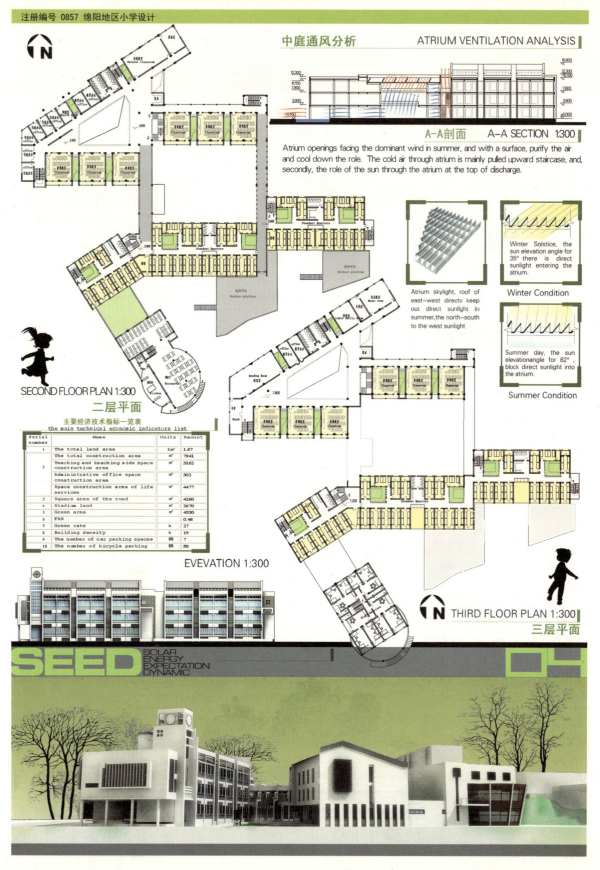

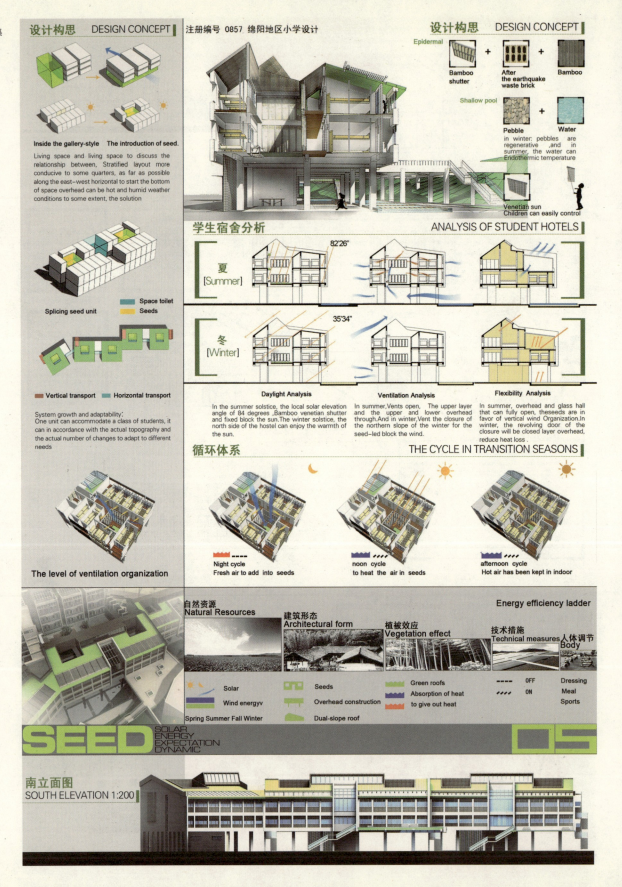

注册编号 0857 绵阳地区小学设计

空气集热器 | AIR COLLECTOR

Solar air heater core collector components by double openings of single-layer glass wall-type collector (1) and its part of the Globoidal axial extension (3), with matching corresponding endothermic layer (2) composition, also has the accessories: the gas import and export in conjunction boxes (4) and (6) and stenting (8), insulation layer (9) and gas flow system (10); it can facilitate the use of solar energy heating the air, using in the regulation of the corresponding facilities (11) of temperature, thus saving a lot of energy, and simple structure, easy to use, making installation easy and low cost, long life.

沼气利用 | BIOGAS

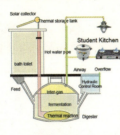

Winter and spring cold season, when the outdoor Temperatures below 15 degrees, open the hot water Control valves, so that heat is stored in the storage tank
Solar hot water, into the methane-generating pits at the end of Department of heat exchanger plates, and fermentation fermentation tanks Raw materials for heat exchange, thereby enhancing the Biogas fermentation temperature. The hot summer season Festival to close the water control valves, hot water is not Re-entering the methane-generating pits, water stored in the storage Box to provide students with hot water

雨水收集 | CULTIVATION PLAYER

SEED
SOLAR ENERGY EXPECTATION DYNAMIC 05

DOUBLE-INSULATING GLASS | 双层中空玻璃

Glass used hollow glass, its most important advantages of heat insulation, thermal insulation, both anti-Condensation, noise and other properties. It can reduce indoor air convection heating, indoor temperature uniformity, Comfortable environment. Heating and air conditioning in the use of environmental Pu, if Insulating glass can be installed to save energy around 30%, which calculated Insulating glass than the single-layer glass increased investment costs in the 5 years left Right from the heating will be able to reduce energy consumptionbyair-conditioning savings in the branch Recovered out.

REFUSE TO USE AFTER THE EARTHQUAKE | 震后废料利用

Post-disaster construction waste recycling renewable brick production the whole process only five steps: the initial clean-up sorting out the recyclable steel; broken into aggregate; screening aggregate formation of qualified building materials; add cement, fly ash, etc. Kevin pressure molding. During the design of the full use of such material, the realization of the 3R materials (Reduce, Reuse, Recycle) higher utilization rate.

GREEN ROOFS | 种植屋面

Planted roof moss is a type of vegetation To the top of the thermal insulation purposes, and Traditional green roof comparison, Reduce the load of the roof structure. And Insulation is still good
Filtering layer (non-woven 1cm) Drainage layer (ceramsite 3-5cm)
EPDM waterproof layer of concrete layer Polyurethane foam insulation layer SBS waterproof layer of concrete layer Floor layer

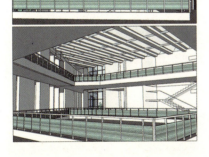

作息时间和太阳能技术 | REST TIME AND SOLAR TECHNOLOGY

时间循环　　　　　　　　　　　场所循环

能量循环　　　　　　　　　　　物质循环

日照分析 | SUNLIGHT ANALYSIS

6 hours
5 hours
4 hours 1 On Spring Vernal Equinox
3 hours 2 On summer Lagopsis
2 hours 3 On winter Big Chill Day

庭院与气候 | COURTYARD AND CLIMATE

Seats can be the local stone, high-back seat width is twice as much, to be used in soil seat. Ride south. Winter soil profile back north, sunlight in the seat. Seat behind the trees can be planted in summer shade from the role.

When the winter sunshine, scattered over the whole body when a person is quiet reading, enjoying their own time. Some students study with homework, share a secret.

When the winter sunshine, scattered over the whole body, the all singing dancing, shouting all day happy years. Become a public place of rest.

Chair form of the use of arc can be used as outdoor teaching areas and student exchange zone. Its arc-shaped chair with the movement of the sun one day, people can at any time to find a place direct sunlight.

[阳光小学 SOLAR ELEMENTARY SCHOOL 1]

二等奖
Second Prize

项目名称：阳光小学
　　　　　Solar Elementary School
作　者：房涛、刘震、李默、杨文江、
　　　　　吴昊、张乐岩
参赛单位：山东建筑大学建筑城规学院

专家点评：
总平面布局简洁合理；采用了雨水收集系统、遮阳板等被动技术；太阳能的应用面积比较大；西南墙的设计很符合当地的建筑风格。
缺点是餐厅建筑跟整体不协调，缺乏与宿舍和教室的联系；空间有些单调。
建议餐厅的颜色风格尽量与总体一致。

The layout is sententious and reasonable. Some of passive technology is adopted such as rain water collection system, sun shades, etc. The floor area utilized with solar energy is bigger. The design of south-west wall is accordant with local architectural style. There are some shortcomings that the canteen building is not harmonious to the whole and the hostel is short of connection with class rooms. Besides, the spaces are a bit monotonous. A suggestion is that the color and style of canteen should be endeavored to be consistent with the whole.

平面上采用当地民居布局形式，形成院落空间。结合当地风向，西侧开敞有利于夏季自然通风，东侧封闭，有利于阻挡寒冷冬季风。
Plane layout use of local residents to form a courtyard space, combined with local wind direction, the west side open for natural ventilation in summer, the east side of the closure will help block the cold winter monsoon.

校门构思/Gate conception
1. 校门提取当地传统民居构架形式，体现地方特色。
Extracting local traditional forms of residential architecture to Embody local characteristics.
2. 门楼高起，引导视线。
Gatehouse high to guide the line of sight.
3. 大门采取梁柱形式，隐喻栋梁。

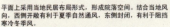

教学楼结合中国传统九宫格形式，形成模块化空间，可以灵活地进行空间组合。
Teaching Building of combining traditional Chinese form of squares to form a modular space, which will have the flexibility to carry out space portfolio.

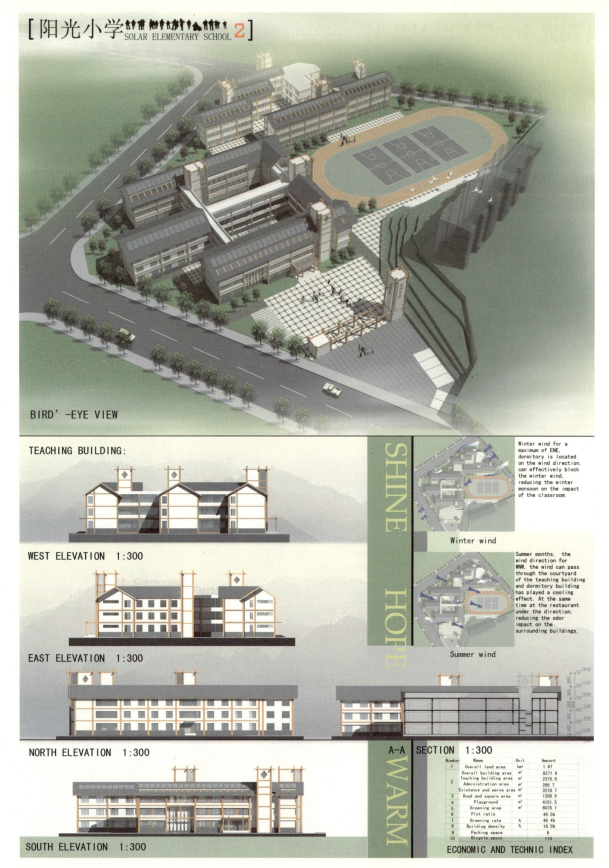

[阳光小学 SOLAR ELEMENTARY SCHOOL 3]

2009 台达杯国际太阳能建筑设计竞赛获奖作品集

FIRST PLAN 1:300

Module Analysis

To content the design to adapt to different use of the functional changes in teaching modules will be modular

Portfolio Analysis function through a combination of different functional areas, to adapt to different bases, and more adaptability to content the design requirements

- Hall
- WC
- Common classroom
- Especial classroom
- Office

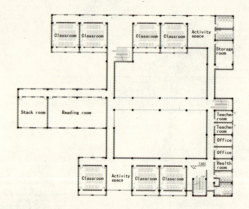

SECOND PLAN 1:300

THIRD PLAN 1:300

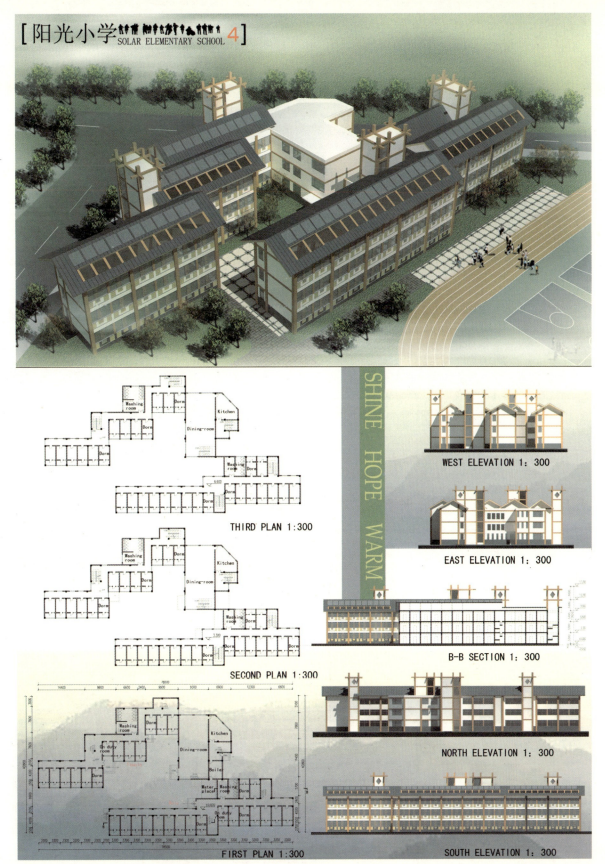

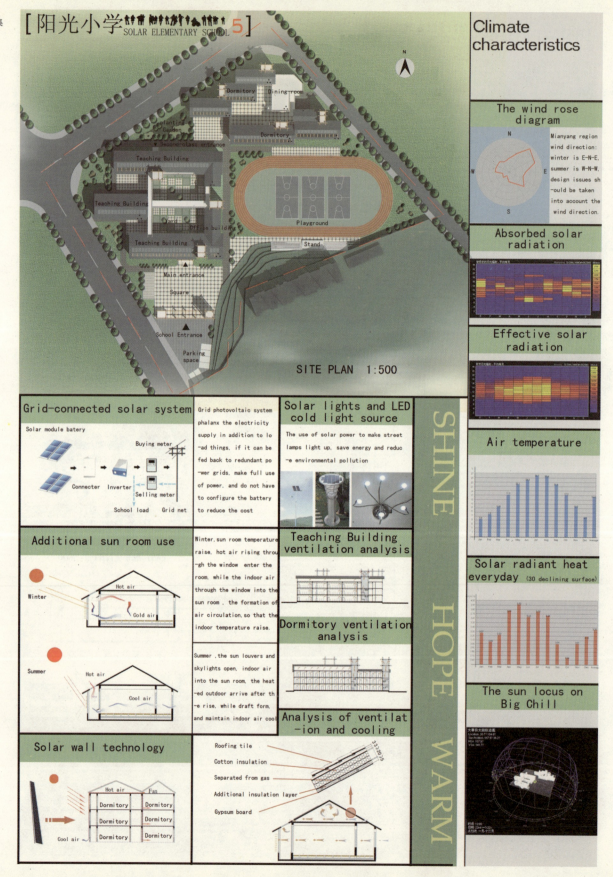

Technology Analysis

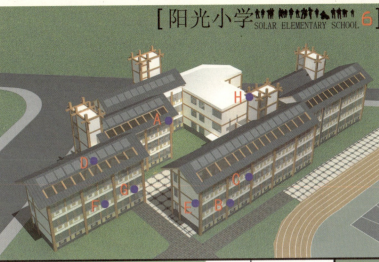

A Rainwater collection system

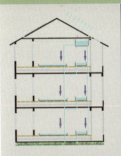

The system can take advantage of rainwater to wash their hands and flushing, while washed hands can also be flushing water, saving water resources.

B Adjustable shade and light analysis (First and second floor)

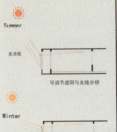

Make use of local bamboo shade adjustable louvers to regulate the indoor temperature in different seasons.

The use of adjustable reflex reflector plate light, adjust the indoor lighting Indoor lighting evenly reflective panels.

C Adjustable shade and light analysis (Third floor)

Because the slope of the roof window, the three-tier roof windows can be used to obtain sunlight, higher utilization of the sun.

Reflective panels also can reflect more sunlight into the room. The room can be more brightness.

E East and West wall insulation design and West wall greening

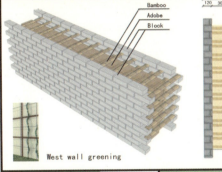

West wall greening

Use of local tradition with thermal escape of the adobe wall, increasing the performance of thermal insulation. Adobe block and bamboo to connect with local materials to increase the overall performance of the wall to reduce the thickness

West wall in the construction of green cultivation Tetrastigma plants, absorb the excessive solar radiation and reduce the external surface temperature, can also be used in wall.

F Low-E Glass

Solar energy 100% → Transmission solar energy 65%

Projects the hot biography loss of heat

In room cold thermal energy 100%

The colloid that can absorb humidity

Air

Low-E film

Outdoor environment   Indoor environment

Four sides the wall glass uses the spatial glass, the heat preservation of the spatial glass is very good. It is utilized widely. Low emissivity glass heatproof radiation, has good heat preservation effect.

G Broken Bridge Aluminum

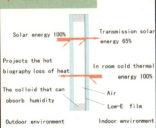

Heat insulation bar   Aluminium alloy window

Windows of this form of heat transfer is very low, can effectively prevent the heat in the windows on both sides of the heat exchange, has played the role of insulation, the application of this technology in the future will be more extensive.

D Solar Water Heating System

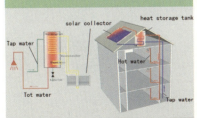

Through the solar collectors, heat medium heat, heat medium in the tank of cold water in the heating cycle, while the use of electric auxiliary heating in order to ensure cloudy and rain and snow sunshine adverse circumstances, such as water supply

H Sewage source heat pump system

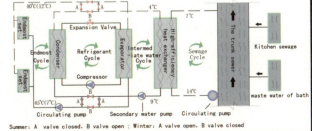

Summer: A valve closed, B valve open ; Winter: A valve open, B valve closed

Sewage source heat pump system can make full use of sewage and heat, including kitchen and bath water. The system not only reduces the heat loss and save energy.

SHINE HOPE WARM

二等奖
Second Prize

项目名称：时光游廊
The Sunshine Corridor of Time

作　者：韩超、许杰、梁叶、马邈、孟光、陈佳敏

参赛单位：山东建筑大学建筑城规学院

专家点评：

建筑设计合理，符合学校总体布局；作品中可以清晰地看到太阳能是怎样设计和利用的，不仅有光热的利用，还有光伏的利用，对学生来说也是知识的普及；利用了地表辐射采暖系统、特隆布墙、反射板等很多太阳能技术；并且考虑了整个建筑的经济性。

也有一些缺点，比如光伏板的位置没有明确标出；热水系统应该是连在一起形成一个系统，而不应该是分开的。

The design is reasonable and meets the requirement of the school layout. In this scheme it is clear to see how solar energy is designed and utilized not only on light-heating but photovoltaic system. It is also a popularization of knowledge for the students. A good many of solar technology are adopted such as heating system of earth surface radiation, trombe wall, baffle-board, etc. Also the economy of the whole building is well considered.

Some shortcomings, for example, the position of PV boards has not been marked out clearly and hot water supply should be connected as a system instead of separate devices.

时光游廊
太阳能小学设计
THE SUNSHINE CORRIDOR OF TIME

GRADE ONE

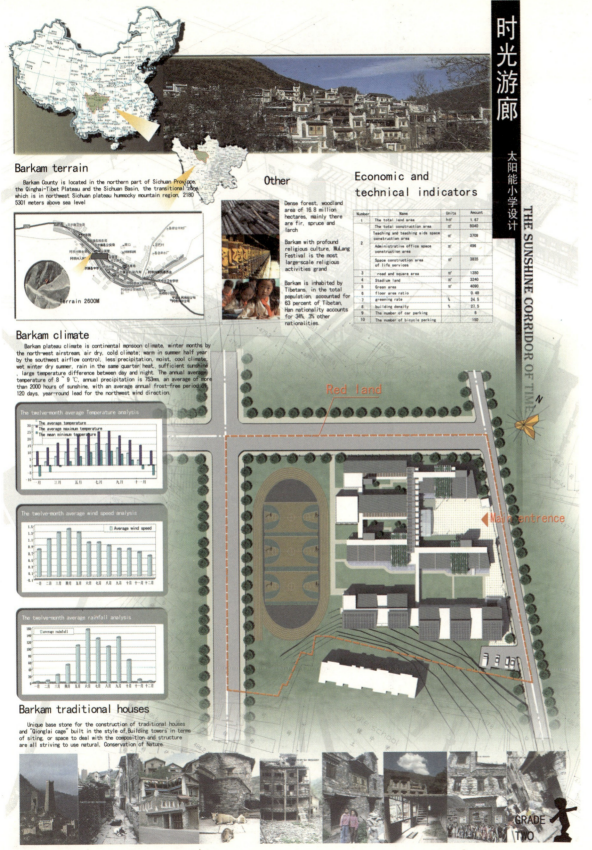

2009 台达杯国际太阳能建筑设计竞赛获奖作品集

时光游廊
太阳能小学设计

THE SUNSHINE CORRIDOR OF TIME

Design Notes
Primary education system is very long, hard and fast learning environment easily tired. So we designed a high, medium and low three classroom groups, linked to rely on the promenade. Students every two years can be transferred to a new study environment. Each group different activities have their own space. The prototype came from the traditional houses in Barkam dry dam, as well as indoor and outdoor space for a combination of ash deformation, an increase of activity space diversity. Barkam the meaning of the Tibetan language are strong flames, promenade as the sun and fire. In building the center, technically making it light and heat generated primary and secondary school students over time and across groups transform the promenade, which represents each time through on growth.

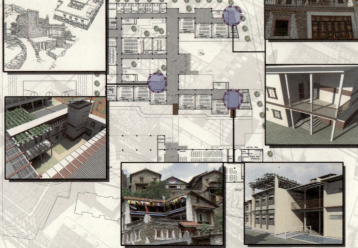

A sunny promenade,
So that the bright young minds to accept baptism.

A recreation of the promenade,
Carries lot of happy childhood memories.

Promenade a time,
Before they are through, witnessed the growth of children.

Physical evolution

Traffice Analysis

- Outside traffic
- Inside traffic
- Transport core

Function Zoning

- The low grades
- The medium grades
- The high grades
- Assisted classroom
- Working area
- Living area

Landscape Analysis

- Landscape node
- Landscape infiltration

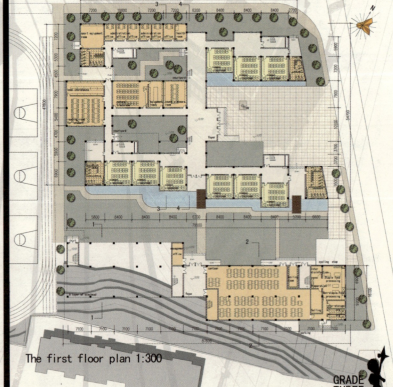

The first floor plan 1:300

GRADE THREE

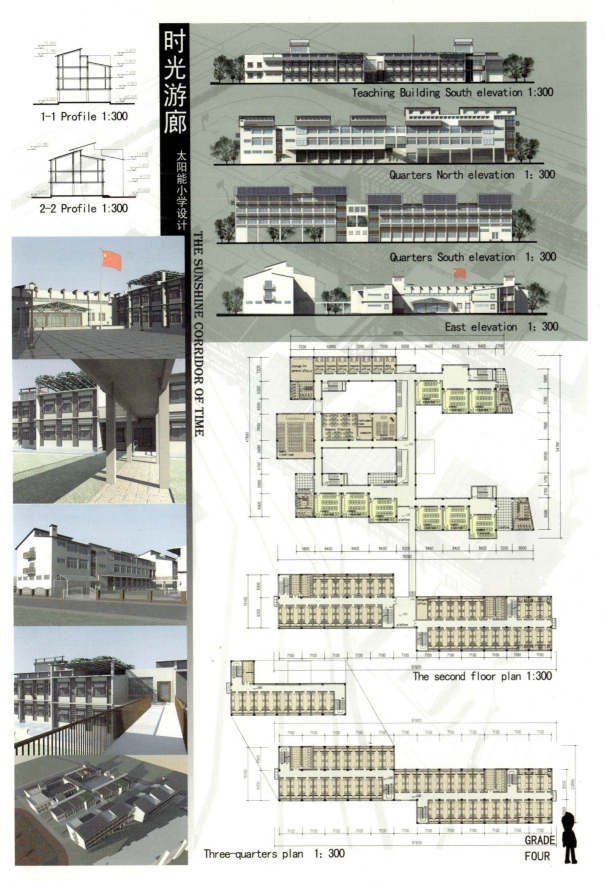

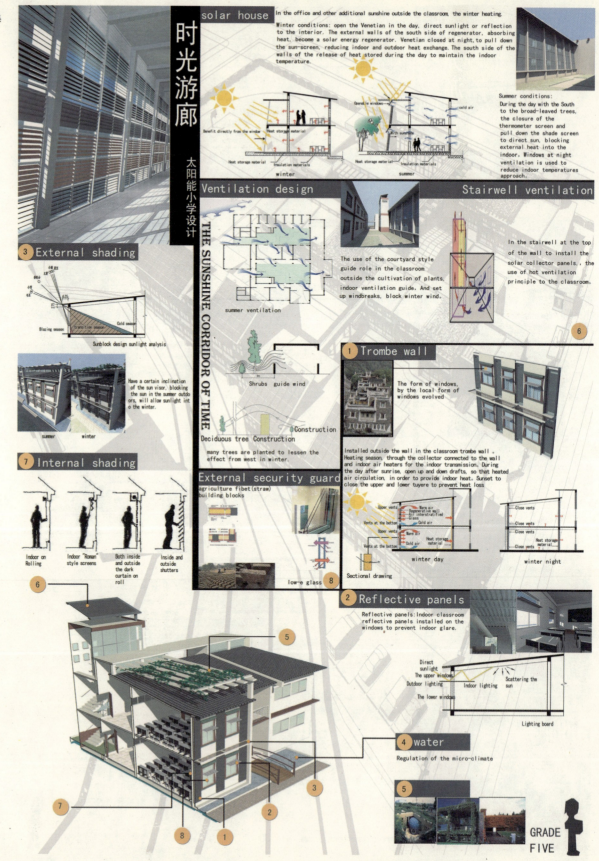

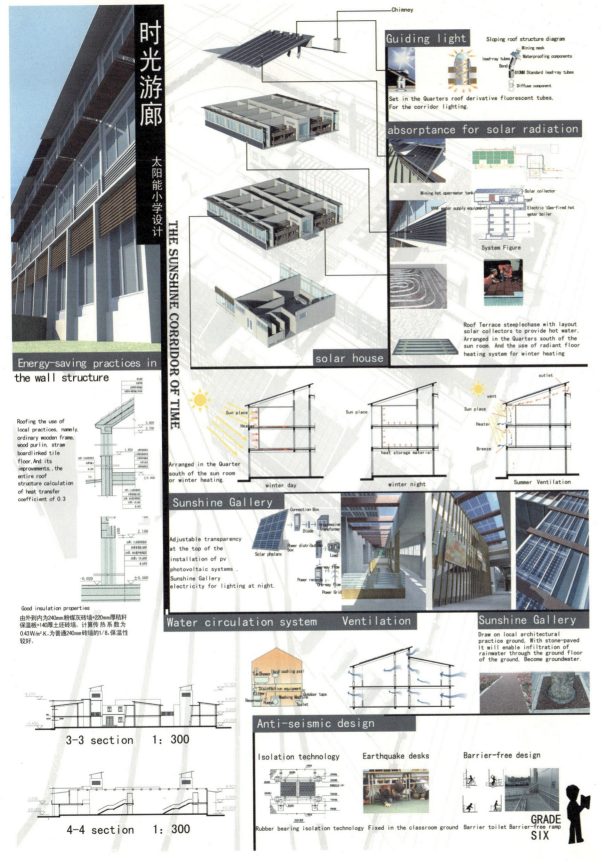

三等奖
Third Prize

项目名称：绵阳阳光小学规划设计
ICNRUS
作　者：班婧、张复昊、李江、胡涛
参赛单位：西安交通大学人居学院

专家点评：

该方案规划布局较为合理，有利于太阳能的利用，功能分区比较明确；在太阳能应用技术与建筑相结合上有创新。缺点是建筑风格与乡村小学不协调，还应该注意方案在工程应用中的可操作性和实用性。

Comparatively the scheme is reasonable on the layout and profitable to the utilization of solar energy. The functional zoning is clear. There is something original on the combination of solar technology application and architecture. The shortcoming is that the architectural style is not harmonious with that of a countryside school. Besides, it should be paid more attention to operability and practicability of engineering applications.

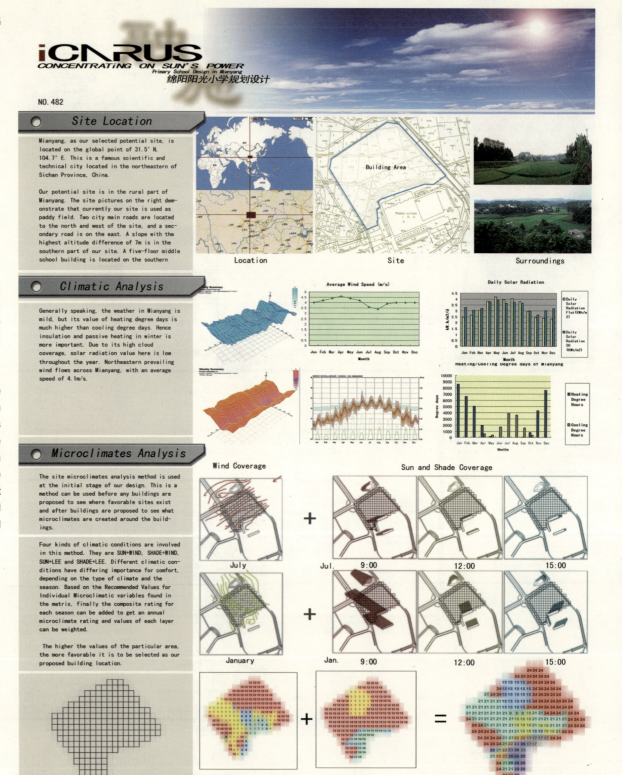

iCARUS
CONCENTRATING ON SUN'S POWER
Primary School Design in Mianyang
绵阳阳光小学规划设计

No. 482

Overall Perspective

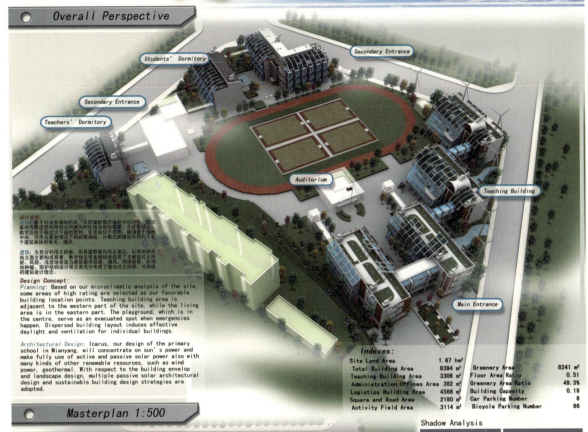

Design Concept:

Planning: Based on our microclimatic analysis of the site, some areas of high rating are selected as our favorable building location points. Teaching building area is adjacent to the western part of the site, while the living area is in the eastern part. The playground, which is in the centre, serve as an evacuated spot when emergencies happen. Dispersed building layout induces effective daylight and ventilation for individual buildings.

Architectural Design: Icarus, our design of the primary school in Mianyang, will concentrate on sun's power and make fully use of active and passive solar power also with many kinds of other renewable resources, such as wind power, geothermal. With respect to the building envelop and landscape design, multiple passive solar architectural design and sustainable building design strategies are adopted.

Indexes:

Site Land Area	1.67 hm²	Greenery Area	8241 m²
Total Building Area	8394 m²	Floor Area Ratio	0.51
Teaching Building Area	3306 m²	Greenery Area Ratio	49.3%
Administration Offices Area	382 m²	Building Capacity	0.19
Logistics Building Area	4566 m²	Car Parking Number	8
Square and Road Area	2180 m²	Bicycle Parking Number	86
Activity Field Area	3114 m²		

Masterplan 1:500

Details:
1. Students' Dormitory
2. Kitchen & Boiling Water Room
3. Canteen for Teachers and Students
4. Teachers' Dormitory
5. Comprehensive Building
6. Teaching Building
7. Auditorium
8. Sport Equipment Room
9. Bio-compost
10. Reserved Raddy Field
11. Natural Science Area
12. Car Parking
13. Bike Parking
14. Artificial Marsh
15. Wind Turbine
16. Sunflower PV
17. Flagpole
18. Secondary School Teaching Building

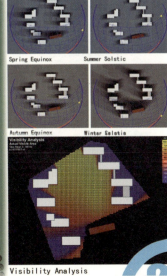

Shadow Analysis — Spring Equinox, Summer Solstic, Autumn Equinox, Winter Solstic

Visibility Analysis

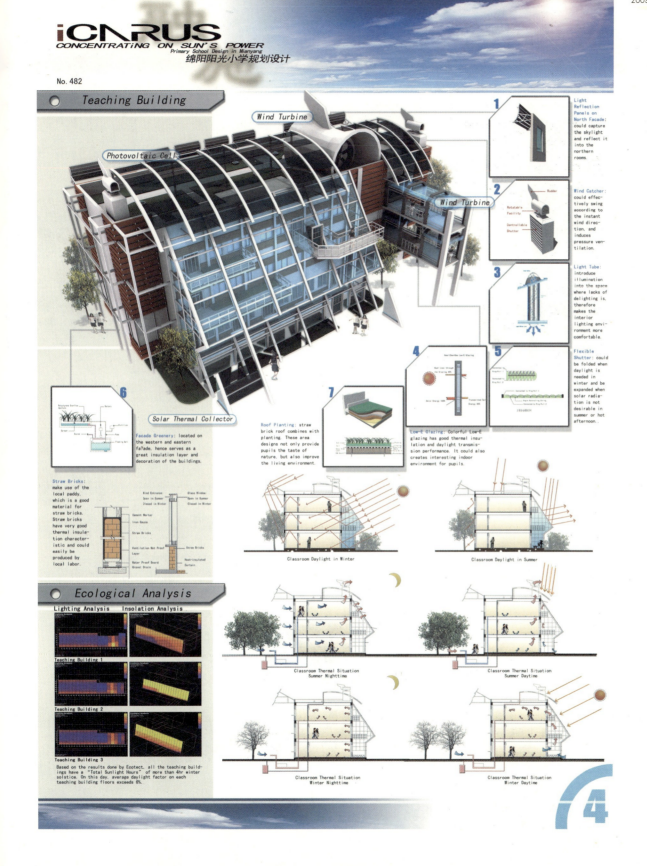

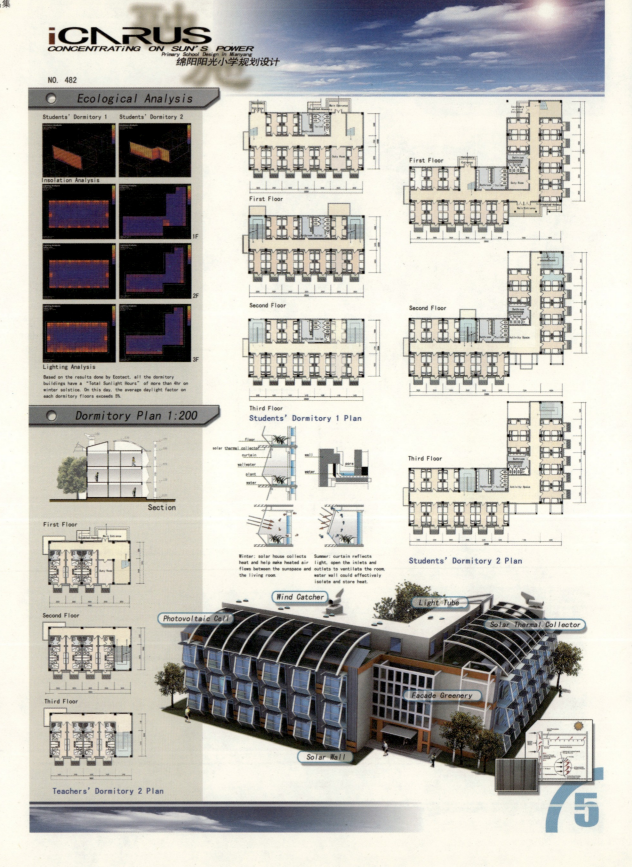

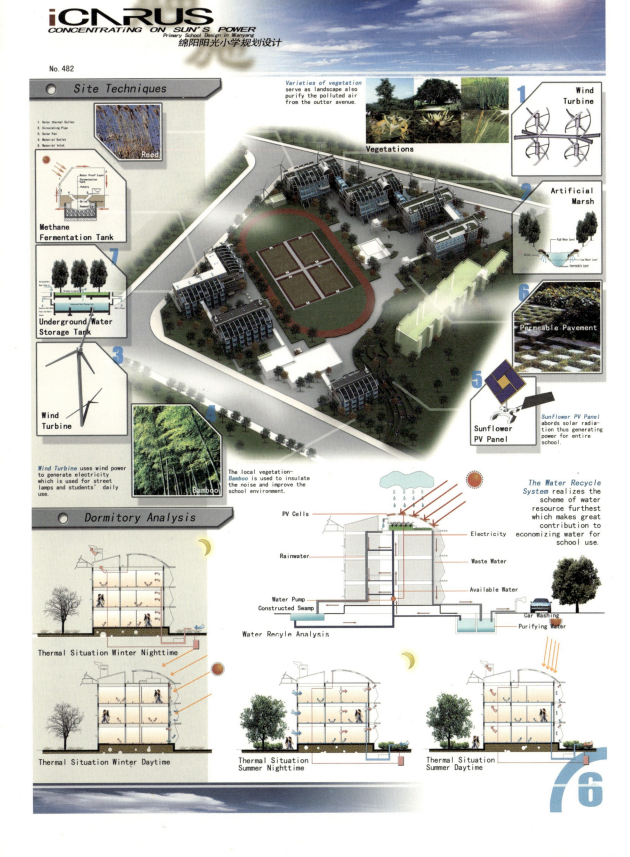

三等奖
Third Prize

项目名称：长屋
　　　　　Long House
作　 者：叶佳明、梁博、杨晓杉
参赛单位：东南大学建筑学院

专家点评：

建筑规划较为合理，功能分区比较明确，有利于太阳能的利用。在自然通风上有一定的创新；但建筑风格和形式不适合四川盆地的建筑风格。

The architectural planning is reasonable comparatively and the functional zoning is clear. They are profitable to the utilization of solar energy. There is something original on natural ventilation. However, the architectural style and form are not suitable to that in Sichuan basin.

Long House
International Solar Building Design Competition
NO. 597

设计说明

我们通过对资料的研究和分析认为，本设计面临的最大问题是如何解决绵阳地区夏季炎热潮湿的气候环境，并同时要兼顾考虑建筑抗震和冬季保温的功能，为学生们造一个舒适的学习生活空间。本设计以此为出发点，试图通过建筑手段来解决这一问题。

Design Concept

The main concept is trying to solution the problem about the hot and wet weather in summer in the mianyang area. And also take the earthquake protection and winter warm keeping into account. Creating the comfortable space for students studying and living is our final target

The Main Economic and Technical Indicators

Gross site area:	16707m²
Gross floor area:	7972m²
Teaching and teaching aids space area:	3012m²
Administrative area:	385m²
Living space area:	4575m²
Road and square area:	1523m²
Playground area:	3564m²
Green area:	1506m²
FAR(excluding playground area):	0.477
Greening rate (excluding playground area):	9.37%
Site coverage:	16.1%
The number of car parking spaces:	7
The number of bicycle parking:	805

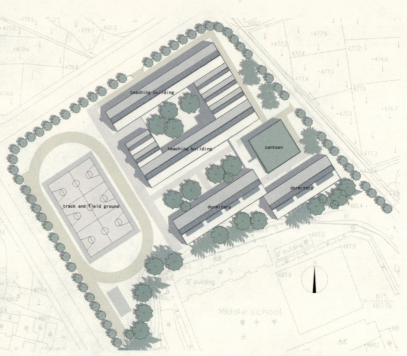

Siteplan 1/500

Perspective

Long House
International Solar Building Design Competition
NO. 597

Plan

Ground Floor 1/300

Model Photo

1. The model for mastplan analysis

We make the sketch model to analyse the mastplan. Because we want to guide the summer wind (from southeast) into the school courtyard, taking the heat away from school, and keeping the whole camps dry.

At the same time, the building position must resist the winter wind (from northeast) effectively.

2. The model for second floor light enviroment analysis

We make this model to do some reseach about the second floor light enviroment, which is use the skylihgt as main light source at the day time. Then we find some problems, which guide us to do some detail design to solution these problems.

Section

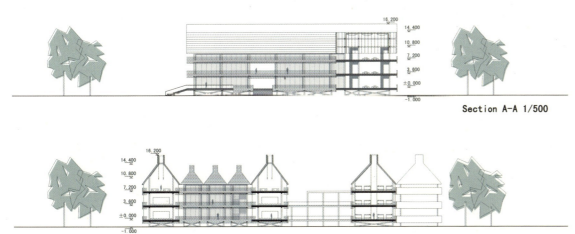

Section A-A 1/500

Section B-B 1/500

Long House
International Solar Building Design Competition
NO. 597

Plan

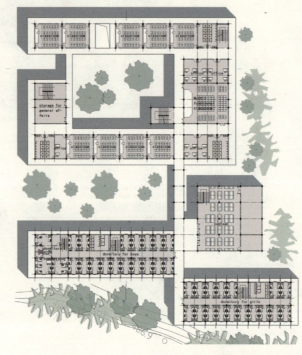

First Floor 1/300

Model Photo

The model for facade design

When we think about the facde design, we want the classroom catch the sunshine as much as possible, and keep the hot outside the classroom. It looks like very conflict, so we just want to find the blance point between these two problems.

The model for facade design

We design the handtrail as an important sunshade element. We mske the sketch model's facade fill with the handtrail, just like the window close by the venetian, Then we delete the handtrail one by one, and check the light enviroment at the same time. In this way we find the handtrails number at last.

Elevation

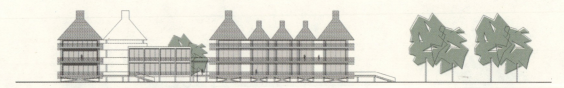

Elevation East 1/300

Elevation South 1/300

Long House
International Solar Building Design Competition

NO. 597

Teaching Building Detail Drawing

Teaching Building Sectional View

Roof Construction
aluminum plates, type 65/400
270mm×120mm×11mm steel truss
airlayer
waterproofing
battens with insulation in between
vapor barrier
steel sheets
steel beam IPE 300mm

facade
aluminum handrail
substructure steel
window, wood and aluminum with triple
6mm insulated glass +12mm cavity + 4mm
insulated glass
transom light

floor construction
25mm oak parquet flooring
60mm screed
40mm acoustic insulation
10mm cement-bound eps
220mm reinforced concrete

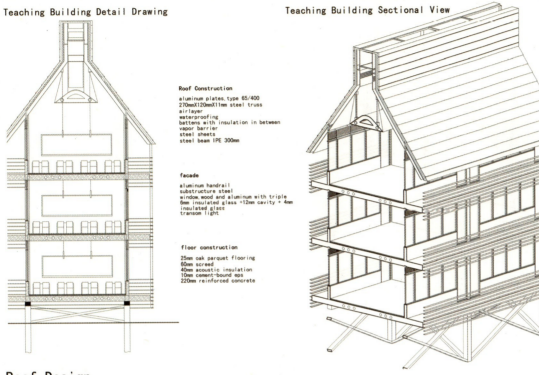

Roof Design

Design Concept

The most problem in Mianyang area is the weather in summer is hot and wet. This roof try to find a way to solution this problem.

1. **Airlayer:** This area is designed to insulation the hot from sunlight irradiation on the roof surface.

2. **Windsail:** This element is designed to guide the cool wind into the roof, and take the hot which is from airlayer and the room to the outside, keeping the indoor space coolness and dry.

3. **Rooflight:** Through the gemels transmission to control the rooflight close and open. So the motivity can be person or electromotor.

4. **Lightreflector:** If there is no light reflector, The sunlight through the rooflight will go into the room directly. The birght area and the dark area will be separated very clear. It is not fit for study obviously. The function for the lightreflector is to reflect the direct sunshine to the roof and go into the indoor space. The light will not so glare as before.

Roof Element
1. airlayer 300mm
2. wind sail
3. roof light
4. gutter
5. light reflctor

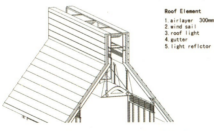

Roof Section

Roof Sectional View

Roof Ventilation

Through the window, door and transom light, The fresh air gets in the second floor. At the same time the hot air comes out with the help of the windsail, air circulation is guarantee

Lightreflector

The lightreflect is a high reflective aluminum plate. At the day time it reflect the sunlight into the room, at night it also can reflect the lighting effectively, can saving the electricity.

Roof Ventilation   Windsail

day

night

Long House
International Solar Building Design Competition

NO. 597

Plan

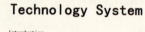

Second Floor 1/300

Technology System

Introduction

1&2. Sun and rain protection is provide by overhanding handrail which are surrounding around the building. The second floor is protected by an overlapping roof giving shelter. The middle floor' shandrail are extended downwards, therefore they supply sun and rain protection for the ground floor.

3. The truss at the bottom of the building provide the protection when the earthquake happen.

4. Vegetaion in southern courtyard allows for heat gain in winter and sun shading in summer

1. Sun Protection

2. Rain Protection

3. Eerthquake Protection

4. Vegetaion in southern coutyard

Elevation

Elevation West 1/300

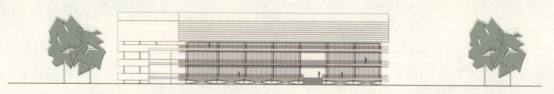

Elevation North 1/300

Long House
International Solar Building Design Competition
NO. 597

Dormitory Detail Drawing

Dormitory Sectional View

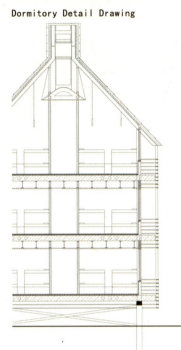

Roof Construction
solar collector
aluminum plates, type 65/400
270mmX120mmX11mm steel truss
airlayer
waterproofing
battens with insulation in between
vapor barrier
steel sheets
steel beam IPE 300mm

facade
aluminum handrail
substructure steel
window, wood and aluminum with triple
6mm insulated glass +12mm cavity + 4mm
insulated glass
transom light

floor construction
suspended ceiling for ventilation
25mm oak parquet flooring
60mm screed
40mm acoustic insulation
30mm cement-bound eps
220mm reinforced concrete

Ventilation System

Teaching Building Roof Sectional View

Teaching Building Ventilation System
rotary venetian fanlight
6mm insulated glass +12mm cavity + 4mm insulated glass
220mm thermal storage wall

Dormitory Ventilation System
suspended ceiling for ventilation
6mm insulated glass +12mm cavity + 4mm insulated glass
220mm thermal storage wall

Dormitory Roof Sectional View

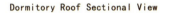
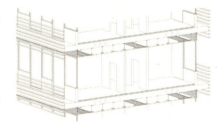

Cross Ventilation for Teaching Building
The facades in each floor can be opened to the north and the south side. Cause of thermal exchange fresh air at the lower part is coming into the room. The air rises because of warming and get out through the fanlight to the outside. Therefore a breeze through the building is secured. Cause of lifting the ground floor fresh air is circling underneath the building for coooling and drying.

Cross Ventilation for Dormitory
This is the same theory for doemitory ventilation as teaching building. Because of the different plan design, we design the suspended ceiling for ventilation. We set the vent in each room and corridor, the warmed air get out through the suspended ceiling.

Teaching Building Ventilation   **Dormiyory Ventilation**   **Lifting Ground Floor Ventilation**

Solar Collect System

Solar Collect Function

There is a solar collector on the top of the roof for warming up the water. (Detail about collector can be seen in the dormitory sectional view). When the weather is cloudy or raining, the water can be warmed up by the boiler's heat (from the canteens).

love & Sunshine & Hope
Rural Sunshine Primary School Design in Mianyang

ID: 638 — INTRODUCION NO.1

三等奖
Third Prize

项目名称：爱 & 阳光 & 希望
Love & Sunshine & Hope

作　者：罗智星、谢栋、安赞刚、李琎、
　　　　宋利伟、柯铠、陈晨

参赛单位：西安建筑科技大学

专家点评：
作品整体布局基本满足设计要求，功能分区也较明确；建筑单体在太阳能利用方面设计合理，采用直接收益窗和太阳能烟囱以及热水集热器等技术；同时还考虑了沼气雨水等其他可再生能源的收集和利用；具有较好的可操作性。

The master layout may meet the requirement of design basically. The functional zoning is clear, too. The design about utilization of solar energy for buildings is reasonable such as directly benefiting windows, solar chimney, solar collectors for hot water, etc. Also the collection and utilization of rain water, biogas and other renewable energies have been considered. The scheme has a good operability.

BACKGROUND

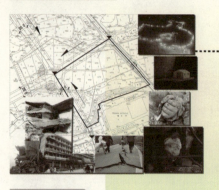

2008年5月12日14时28分 → 2008年6月25日 阳光小学竞赛开始 → 2009年3月 阳光小学方案确定 → 2009年6月 阳光小学动工 → 2010年春季 阳光小学落成

CONCEPT

Security:
Because of 5.12 earthquake, 6 million people were killed and more than 30,000 people were missing. This huge disaster makes us to focus primarily on Security in our design.

suitability:
We emphasize in the design of technical suitability. We believe that suit to the actual situation, adapting to local climatic conditions, finding a balance between the cost and comfort are the most important thing.

Happiness:
These pupils who have experienced the disaster. One of the most important things for us to do is giving them happiness and creating happy spaces to help them learning in the comfortable environment.

CHALLENGE

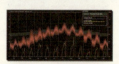
Daily Average

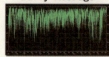
Relative Humdity

Direct Radiation

Prevailing Winds

Weather Challenges
1. Solar energy resource is short. How to use the limited solar energy to offer us hot water and heating in winter?
2. Humidity level is high all the year around.
3. Wind speed is low all the year around. How to design natural ventilation buildings, which providing a comfortable space as well as ensuring damp-proof of the building?

Other Challenges
1. Because it is a rural primary school design, budget is very limit. How to find a balance between the cost and comfort?
2. The way of Sichuan has its own particularity. How to design, making the construction keeps with local atmosphere?
3. The site is very small; the shape is strange; the number of layers is limited. How to improve the design so that we can not only meet the design schedule but also make our design perfect?

ECOLOGICAL UTILIZATIONS

被动式太阳能策略
Passive Solar Energy Utilizations
- 太阳墙系统 Solar Wall System
- 太阳能集热器阵列 Solar Collector Array
- 集热蓄热墙 Thermal Storage Wall
- 直接受益式窗 Directly Beneficial Window
- 太阳能烟囱 Solar Chimney
- 斜屋顶空腔保温 Sloping Roof Heat Insulating Layer

主动式太阳能策略
Active Solar Energy Utilizations
- 太阳能热泵供热和制冷 Solar Heat Pump Heat & Colling

其他生态策略
Other Ecological Utilizations
- 机械通风 Mechanical ventilation
- 屋顶绿化 Roof Planting
- 保温外墙 Thermal Insulation Wall
- 外窗遮阳 Sunshading window
- 密肋复合墙 Multi-ribbed Composite Wall
- 垂直绿化 Vertical Planting
- 生态水池 Ecological Pool
- 雨水收集系统 Rain Water Collection System
- 生态园地 Ecological Garden

2009 DELTA CUP-INTERNATIONAL SOLAR BUILDING DESIGN COMPETITION

love & Sunshine & Hope

Rural Sunshine Primary School Design in Mianyang

PLANNING NO.2

ID:638

2009 台达杯国际太阳能建筑设计竞赛获奖作品集

SHADOW ANALYSE

Shadow of Site 22th, Mar.

Shadow of Site 22th, Jun.

Shadow of Site 22th, Sept.

Shadow of Site 22th, Dec.

We can see from the site shadow analysis, planting sciophiles on the north of the buildings is suitable. And planting heliophilous plant on the south and east of the buildings is suitable, as well as student playing space.

Shadow of Buildings 22th, Mar.

Shadow of Buildings 22th, Jun.

Shadow of Buildings 22th, Sept.

Shadow of Buildings 22th, Dec.

We can see from the site shadow analysis, although the site planning is compact, the building obstruction between the buildings is fully meeting the requirements of sun exposure in winter. At the same time, solar energy can be use well by the three constructions.

WIND ANALYSE

Summer Wind Velocity Analysis

Mianyang is prevailing northwest wind in summer. As we can see from the chart, venue planning is propitious to guide wind in summer. It can also raise the wind speed.

Winter Wind Velocity Analysis(1)

Mianyang is prevailing northeasterly wind in winter. As we can see from the chart, most of the dormitory obstructs most of the wind. But wind guiding is obvious between the dormitory and classroom building.

Winter Wind Velocity Analysis(2)

We can plant some evergreen trees between the dormitory and teaching building. It can reduce the impact on the site obviously.

PLANNING ANALYSE

透水地面
不透水地面
水面

透水地面分析

Through the chart of annual solar radiation analysis, it is certain that we plant sciophiles in the low exposure area, and plant heliophilous plant in the high exposure area.

Annual solar irradiance Analysis

DESIGN NOTE

This program site is located in Mianyang of Sichuan province, latitude 31.5°, longitude 104.7°, measured at 472 meters above sea level. It belongs to China's Thermal Division of the hot summer and cold winter areas. Therefore, in this design, we take full account of the natural ventilation in the summer and thermal comfort without coal-fired heating in winter of the primary school building.

The construction site, which is located around Mianyang, surrounded by farmland and mountains. Town secondary school is at the south of the construction site. The other three borders are next to three rural roads of different level. Main entrance sets on the west trunk in order to make a good series relation with the secondary school, and the children who is living in the settlement surround the site can go to schoole easier. Secondary entrance, which is set on the north trunk, make students more convenient to go to dormitory and playground. Basketball field, volleyball field and track and field ground with 200 meters lane are put in the east of the site synthetically. Natural slope in the south of the site have been kept completely. And we clear up the green area on the site and turn it into outdoor natural science garden.

Master Plan 1:500

Main Technical & Economic Indicators:

1.	Total land area:	1.67 ha.
2.	The total construction area:	7696m²
	Teaching area:	2996m²
	Administrative office area:	402m²
	Service space construction area:	4298m²
3.	Road and square area:	2703m²
4.	Stadium site:	5600m²
5.	Green area:	3512m²
6.	Floor area ratio:	0.876
7.	Greening rate:	40%
8.	Building density:	29.2%
9.	Car parking number:	6
10.	Bicycle parking number:	500

2009 DELTA CUP-INTERNATIONAL SOLAR BUILDING DESIGN COMPETITION

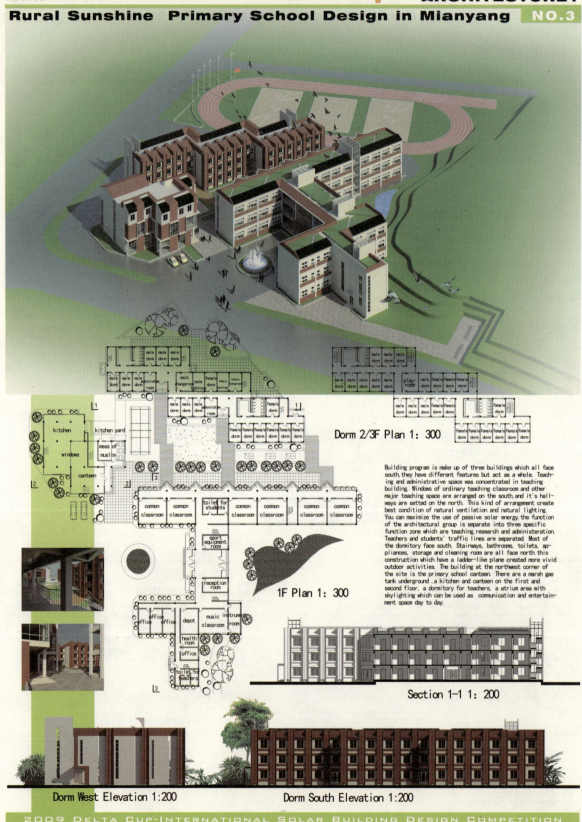

love & Sunshine & Hope

Rural Sunshine Primary School Design in Mianyang

ID: 638 — ARCHITECTURE 2 — NO.4

Main Entrance Perspective

Section 2-2 1:200

Canteen South Elevation 1:200

Canteen 2F Plan 1:300

Canteen 3F Plan 1:300

Teaching Building 2F Plan 1:300

Section 3-3 1:200

Teaching Building North Elevation 1:200

Teaching Building 3F Plan 1:300

Teaching Building West Elevation 1:200

Teaching Building South Elevation 1:200

2009 DELTA CUP-INTERNATIONAL SOLAR BUILDING DESIGN COMPETITION

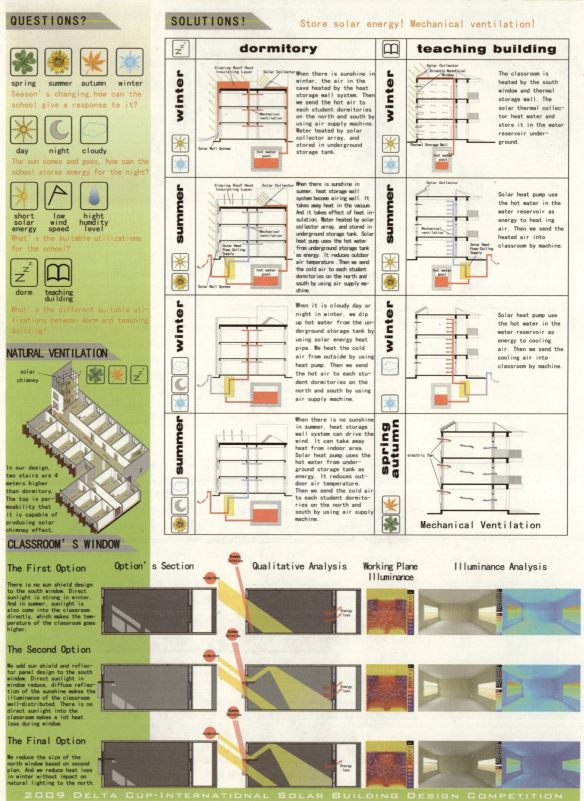

love & Sunshine & Hope
Rural Sunshine Primary School Design in Mianyang
SECURITY & ECOLOGY — NO.6

ID:638

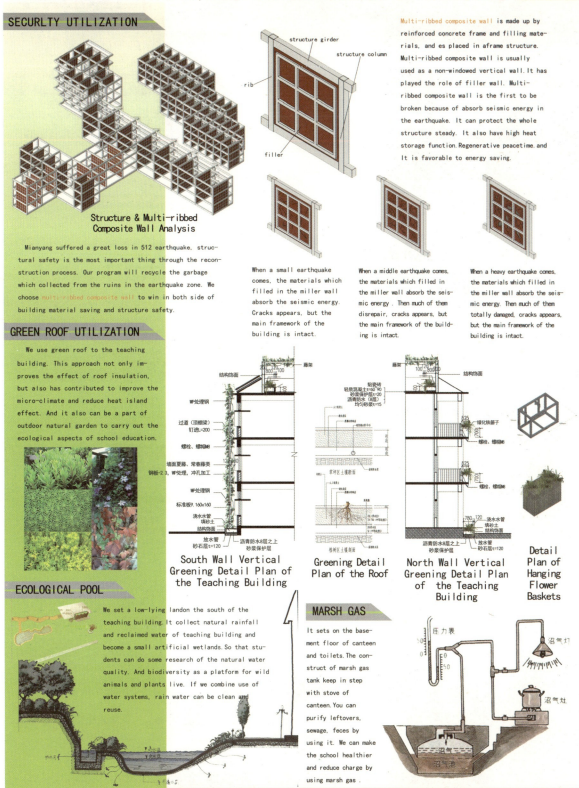

SECURITY UTILIZATION

Structure & Multi-ribbed Composite Wall Analysis

Mianyang suffered a great loss in 512 earthquake, structural safety is the most important thing through the reconstruction process. Our program will recycle the garbage which collected from the ruins in the earthquake zone. We choose multi-ribbed composite wall to win in both side of building material saving and structure safety.

Multi-ribbed composite wall is made up by reinforced concrete frame and filling materials, and es placed in aframe structure. Multi-ribbed composite wall is usually used as a non-windowed vertical wall. It has played the role of filler wall. Multi-ribbed composite wall is the first to be broken because of absorb seismic energy in the earthquake. It can protect the whole structure steady. It also have high heat storage function. Regenerative peacetime. and It is favorable to energy saving.

When a small earthquake comes, the materials which filled in the miller wall absorb the seismic energy. Cracks appears, but the main framework of the building is intact.

When a middle earthquake comes, the materials which filled in the miller wall absorb the seismic energy. Then much of them disrepair, cracks appears, but the main framework of the building is intact.

When a heavy earthquake comes, the materials which filled in the miller wall absorb the seismic energy. Then much of them totally damaged, cracks appears, but the main framework of the building is intact.

GREEN ROOF UTILIZATION

We use green roof to the teaching building. This approach not only improves the effect of roof insulation, but also has contributed to improve the micro-climate and reduce heat island effect. And it also can be a part of outdoor natural garden to carry out the ecological aspects of school education.

South Wall Vertical Greening Detail Plan of the Teaching Building

Greening Detail Plan of the Roof

North Wall Vertical Greening Detail Plan of the Teaching Building

Detail Plan of Hanging Flower Baskets

ECOLOGICAL POOL

We set a low-lying land on the south of the teaching building. It collect natural rainfall and reclaimed water of teaching building and become a small artificial wetlands. So that students can do some research of the natural water quality. And biodiversity as a platform for wild animals and plants live. If we combine use of water systems, rain water can be clean and reuse.

MARSH GAS

It sets on the basement floor of canteen and toilets. The construct of marsh gas tank keep in step with stove of canteen. You can purify leftovers, sewage, feces by using it. We can make the school healthier and reduce charge by using marsh gas.

2009 DELTA CUP·INTERNATIONAL SOLAR BUILDING DESIGN COMPETITION

三等奖
Third Prize

项目名称：普适的地域性
　　　　　Universal Regionality
作　　者：陈宇、吴维聪、王翔、
　　　　　苏岩苁、程冠华
参赛单位：同济大学建筑与城市规划学院

专家点评：
建筑规划布局较为合理；有利于太阳能的利用和自然通风，具有较好的可操作性；但下沉式广场疏散高度超过两层，应注意设计相关要求。

The architectural planning and layout is reasonable comparatively and profitable to the utilization of solar energy and natural ventilation. The scheme has a good operability. However the height for evacuation of the subsidence-type square is more than two stories, it should be paid more attention to relevant requirements concerning building design.

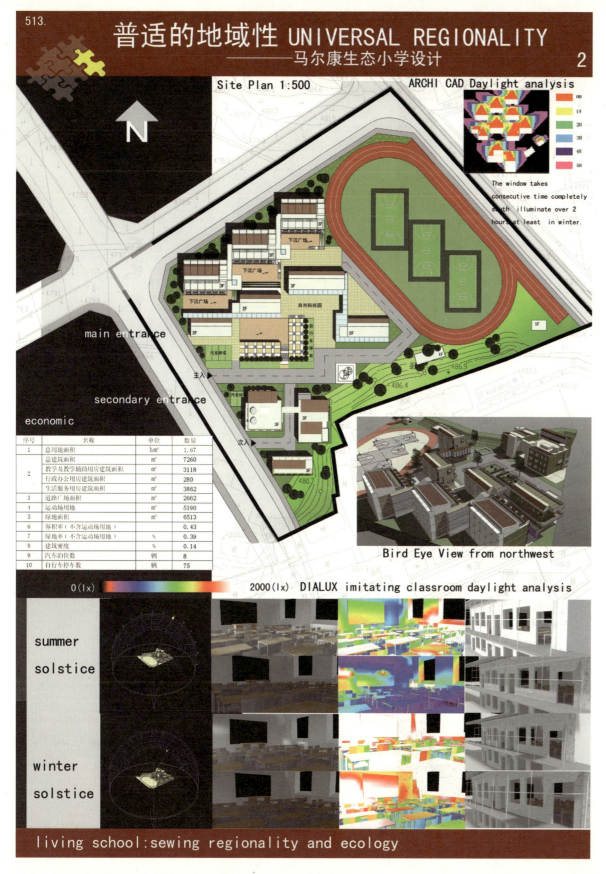

普适的地域性 UNIVERSAL REGIONALITY
——马尔康生态小学设计

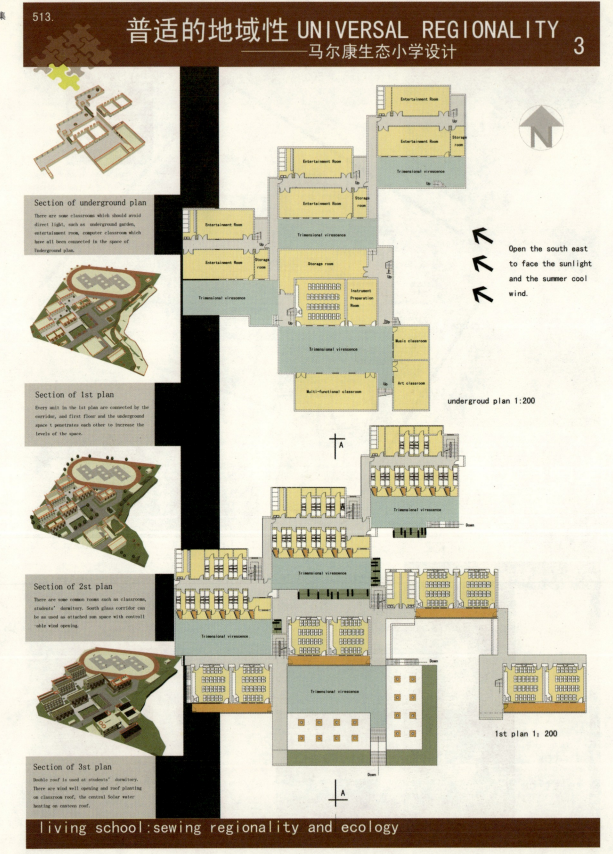

Section of underground plan
There are some classrooms which should avoid direct light, such as underground garden, entertainment room, computer classroom which have all been connected in the space of Underground plan.

Section of 1st plan
Every unit in the 1st plan are connected by the corridor, and first floor and the underground space t penetrates each other to increase the levels of the space.

Section of 2st plan
There are some common rooms such as classrooms, students' dormitory. South glass corridor can be as used as attached sun space with controllable wind opening.

Section of 3st plan
Double roof is used at students' dormitory. There are wind well opening and roof planting on classroom roof, the central Solar water heating on canteen roof.

Open the south east to face the sunlight and the summer cool wind.

undergroud plan 1:200

1st plan 1:200

living school: sewing regionality and ecology

普适的地域性 UNIVERSAL REGIONALITY
马尔康生态小学设计

living school: sewing regionality and ecology

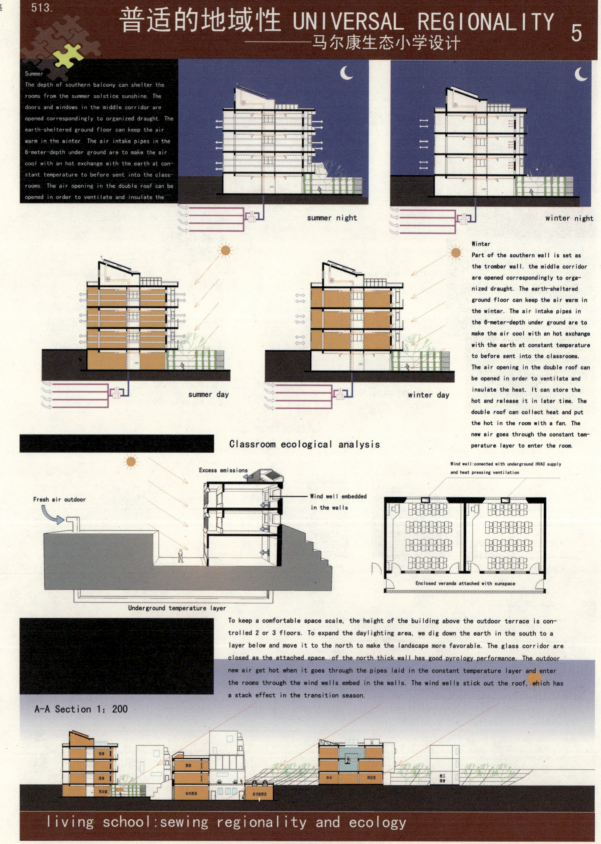

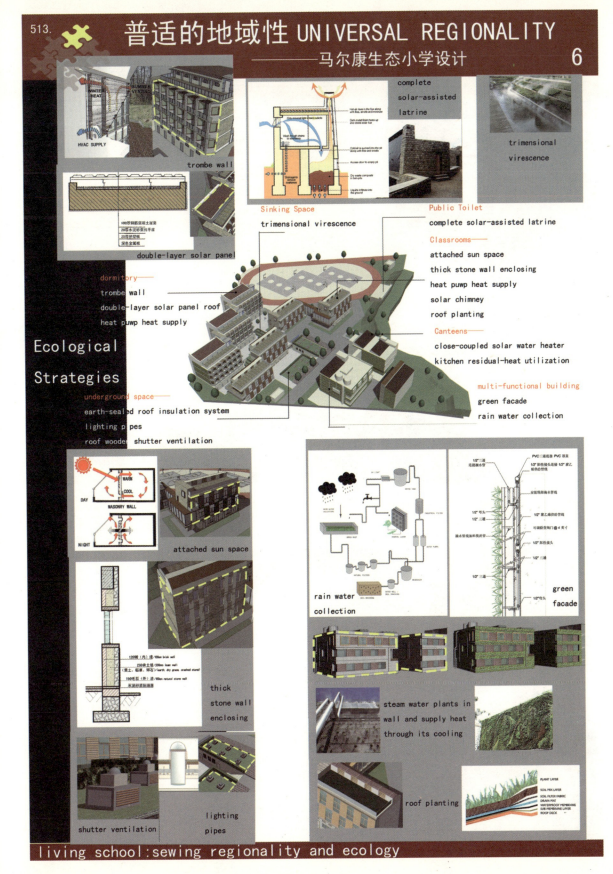

三等奖
Third Prize

项目名称：舞动的藏袍
　　　　　Flowing Tibet Robe
作　　者：何媛、施洁莹
参赛单位：重庆大学建筑城规学院

专家点评：

方案符合竞赛要求，规划布局较合理，建筑主朝向有利于太阳能利用，建筑里面设计有民族特色，具有较好的可操作性。但应注意一层裙房对自然通风和室内庭院日照的影响，以及运动场对教学区的影响。

The scheme meets the requirement of the competition. The layout is reasonable. Main orientation of the buildings is profitable to the utilization of solar energy. The building design has national feature and good operability. However it should be paid more attention to that the annex building of the first floor may affect natural ventilation and sunshine of inner courtyard and the playground may disturb teaching area.

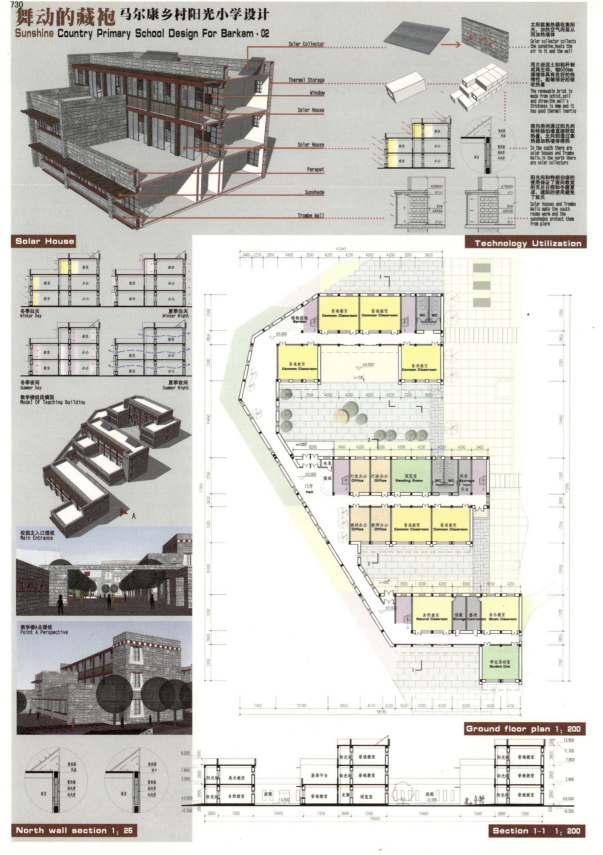

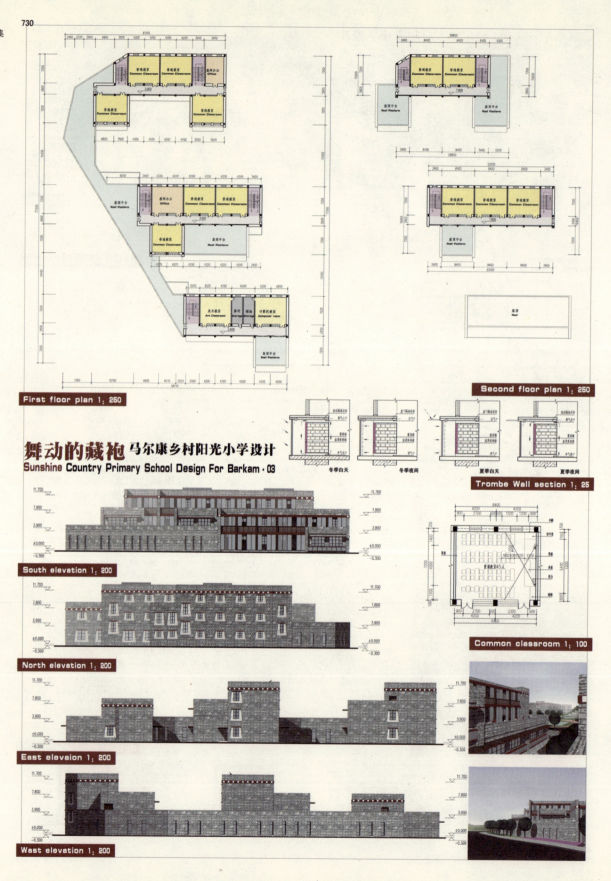

舞动的藏袍 马尔康乡村阳光小学设计
Sunshine Country Primary School Design For Barkam · 03

三等奖
Third Prize

项目名称：气候与塑形
　　　　　Forms Follows Climate
作　者：张振华、范臻
参赛单位：重庆大学建筑城规学院

专家点评：

规划分区明确，布局较合理，建筑单体设计有利于被动太阳能利用；公共活动空间与交通空间有机结合，但应注意建筑朝向应布置在这一地区最佳朝向，即南——南偏西30°，以及注意教学区与餐厅之间的疏散问题。

The planning and zoning are clear and the layout is reasonable. The design of buildings is profitable to the utilization of passive solar energy. The public activity spaces have a good combination with traffic spaces. However it should be paid more attention to the best orientation of the building arrangement in this district i.e. south – 30° West South and the evacuation problem between the canteen and teaching area as well.

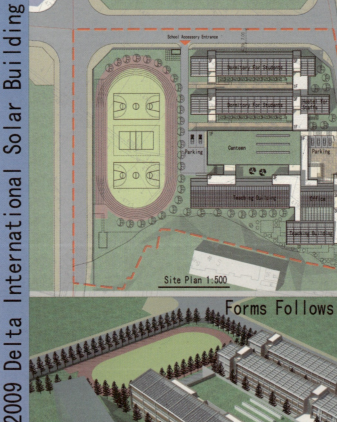

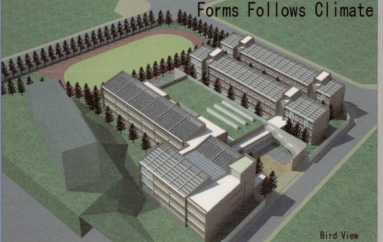

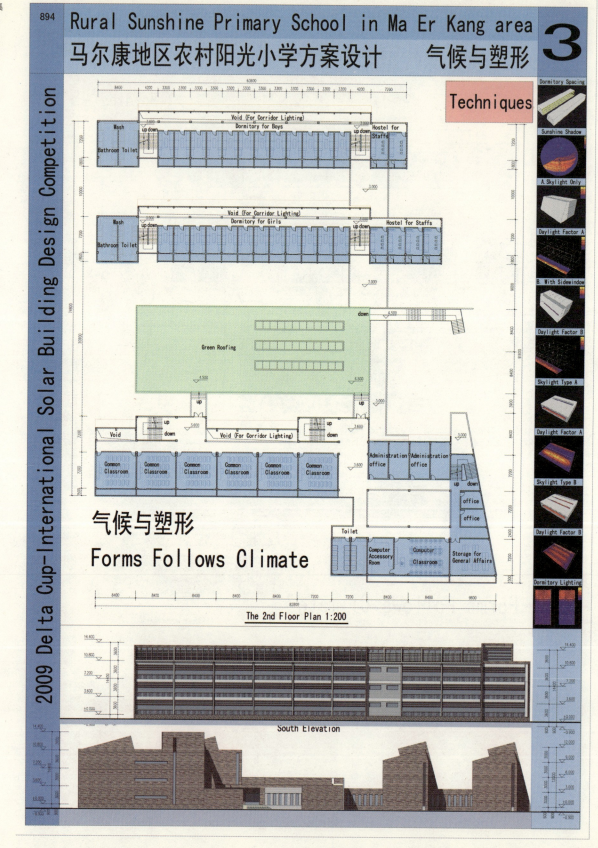

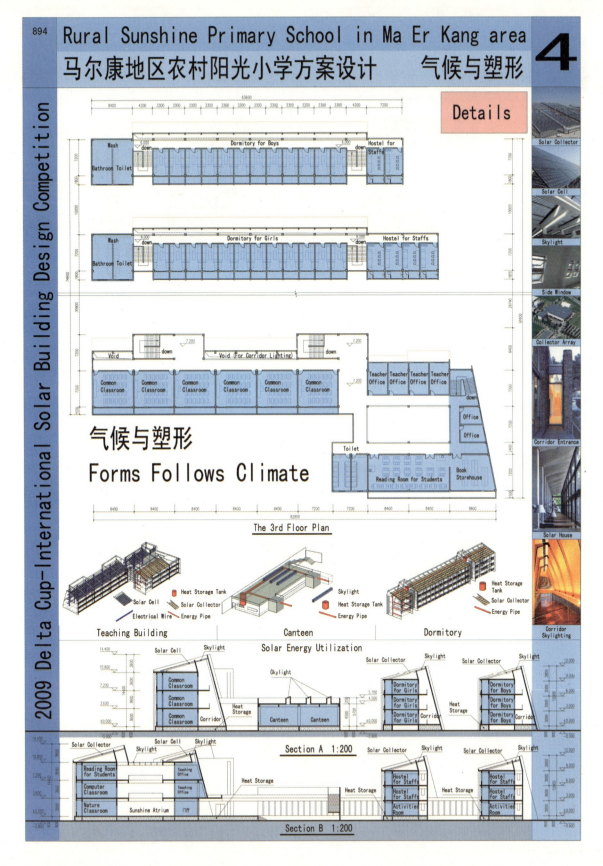

优秀奖
Honorable Mention Prize

项目名称：光之墨迹
　　　　　Sunshine, Ink Marks

作　　者：刘祖国、梅琬菲、文雅

参赛单位：广州大学建筑与城市规划学院，
　　　　　广州市设计院

设计概念
方案设计灵感源于中国传统的书法艺术。"光之墨迹"寓意以光明作笔，空间为墨，在这片满目苍夷的土地上挥写下希望的笔迹。创造最适合于学习与生活的校园环境。

建筑布局
建筑布局规则而灵活。建筑的围合形成良好的空间场所感，创造高效的校园学习，生活空间。同时，建筑的开放空间给予师生舒适的休憩交往空间，体现着人文的设计精神。
四川的气候夏炎冬寒。建筑布局朝向根据当地日照方向，太阳高度角等因素，采取东南３２度３９分的朝向布局。此布局让建筑在夏季避免了强烈的太阳直射。同时，在冬季又可得到柔和的日照取暖，采光，最大限度地减少了建筑的运行能耗。

建筑造型与材料
建筑造型简洁实用，现代主义与地域特色的碰撞达到了简美高效与浓厚地域性的高度融合与统一。
建筑材料选取当地的毛石，竹子，木材等廉价建材，从而降低了工程造价并体现着当地的地域特色。

The Design Concept:
This program's design inspiration is from the art of chinese calligraphy. "The light's painting" expresses that we use light as a pen and space as ink to write down the hope's handwriting in this piece of desolate land.
Create the best campus environment for learning and living.

The construction layout:
The construction layout is inerratic but flexible. Its envelopment form the nicer feeling of space. At the same time, the construction's exoteric space which give teachers and students a comfortable place to relax, incarnate the inspirit of humanism.
Sichuan's weather is hot in summer but cold in winter. The construction's position is under the direction of sunlight, the sun's angle of height and so on, and we position it towards 32° 50' south-east. This design make the construction avoid sun's direct irradiation. Simulanety, it can be warm in winter and reduce the energe consumption the uttermost because of sunlight.

The architectural design and materials:
The architectural design is compact and applied, the modernism and geographical characteristics's impact arrive a high unification and amalgamation with succinctness, efficiency and strong regional. The architectural materials choose the local low-cost stuff such as rubble, bamboo, wood and the like which accordingly debase the project's cost and incarnate its particular characteristics of region.

Bird's-eye view

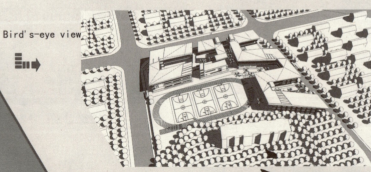

..1　农村阳光小学
　　SUNSHINE HOPE SCHOOL

SUNSHINE, INK MARKS　光之墨迹

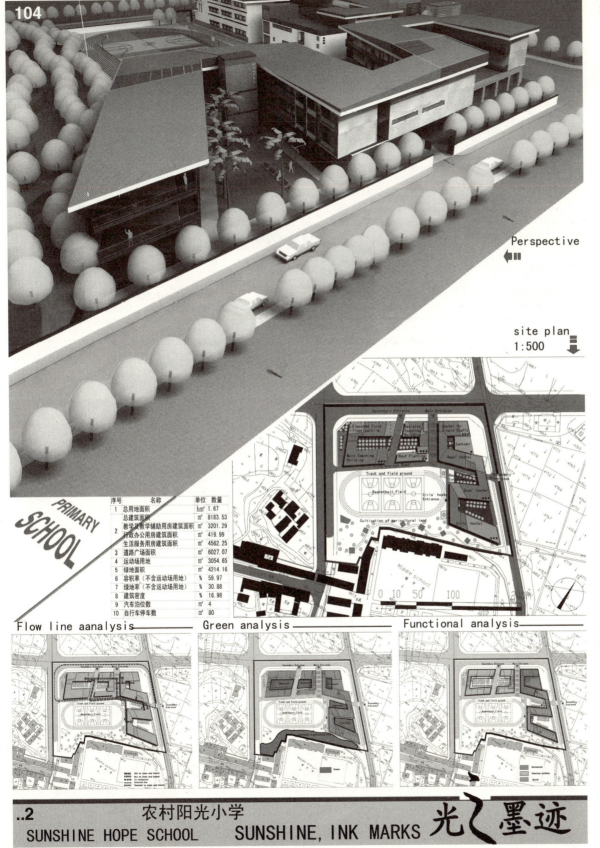

农村阳光小学
SUNSHINE HOPE SCHOOL   SUNSHINE, INK MARKS   光之墨迹

Entrance Perspective

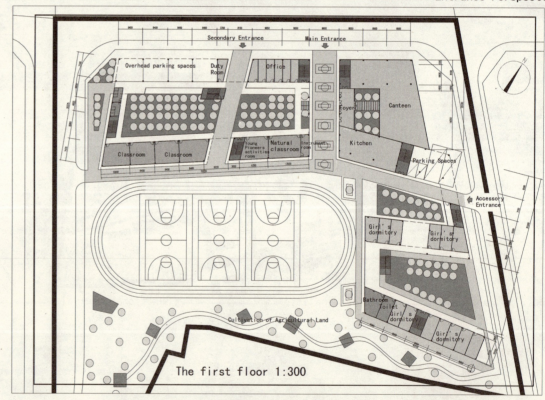

The first floor 1:300

West Facade 1:300

..3 农村阳光小学 光之墨迹
SUNSHINE HOPE SCHOOL　　SUNSHINE, INK MARKS

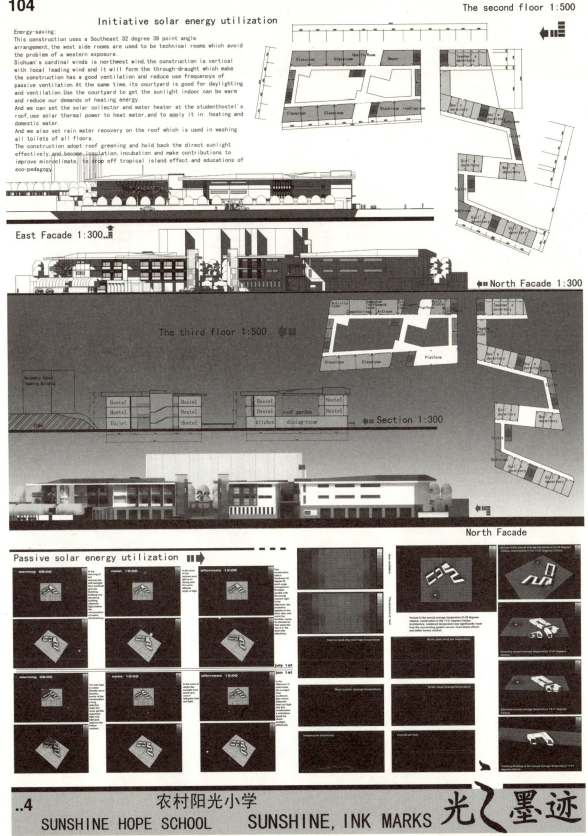

Initiative solar energy utilization

Energy-saving:
This construction uses a Southeast 32 degree 39 point angle arrangement, the west side rooms are used to be technical rooms which avoid the problem of a western exposure.
Sichuan's cardinal winds is northwestern wind, the construction is vertical with local leading wind and it will form the through-draught which make the construction has a good ventilation and reduce use frequencys of passive ventilation. At the same time, its courtyard is good for daylighting and ventilation. Use the courtyard to get the sunlight indoor can be warm and reduce our demands of heating energy.
And we can set the solar collector and water heater at the studenthostel's roof, use solar thermal power to heat water, and to apply it in heating and domestic water.
And we also set rain water recovery on the roof which is used in washing all toilets of all floors.
The construction adopt roof greening and hold back the direct sunlight effectively, and become insulation, incubation and make contributions to improve microclimate, to drop off tropical island effect and educations of eco-pedagogy.

..4 农村阳光小学　　光之墨迹
SUNSHINE HOPE SCHOOL　　SUNSHINE, INK MARKS

绵阳地区阳光小学竞赛设计
MIANYANG SOLAR PRIMARY SCHOOL DESIGN

ENTRY CODE : 115
P 1/4

优秀奖
Honorable Mention Prize

项目名称：绵阳地区阳光小学竞赛设计
Mianyang Solar Primary School Design

作　　者：Xiang WANG、Wenmu TIAN

参赛单位：
Périphériques ARCHITECTES, Paris FRANCE
Ecole nationale Niupérieure architecture de Paris-La villette Paris FRANCE

GENERAL LAYOUT 1/500

ANNUAL SOLAR ANALYSE OF SITE

设计说明 / DESIGN SCHEME

震后学校重建要秉行经济、安全的基本原则。

受震后经济和技术的限制，方案在保证结构简单清晰的原则下，将节能和可持续思想放在设计之初，结合城镇景观和建设功能需求，用简易、廉价的方式，达到优化建筑，乃至整个基地日照方式和能耗。我们避免采用复杂、高技术的设备和构造，而采用调整朝向、合理分配体量和空间、利用自然通风等最基本的设计手段，并配合以廉价的工业建材和材镇便捷可取的生态建材结合的建造方式。我们同时还思考如何将设计融入到城镇景观之中：让基地中心和周边的街道产生间接的视觉联系而不时地制造内和外的对话。这样开放式的机理既增加了未来街道的趣味和可读性，也保留了基地现有地貌的集体记忆。

方案力图减少设计者主观的影响，而尽量由制约设计的城市、经济、文化和技术因素来雕塑其形态。

School rebuilding should follow the fundamental principle of economical and safe.

Because of the economical and technical restriction after the earthquake, our plan first considers energy-saving and sustainable concept, while ensuring the principle of making the structure simple and clear.
Through analyzing the city's landscape and the construction's functional demands, it intends to optimize the sunshine and the energy consumption of the building and the entire site with a simple and low-priced method.
We avoid using complicating and high-technical facilities and structure. Instead some basic design methods are chosen such as modifying its location, separating its blocks and spaces, natural ventilation, utilizing both low-price mechanical material and more accessible ecological local material.
In the mean time, this plan also concentrates on the problem about how to harmonize the building with urban landscape: by creating indirect visual connection between site center and the surrounding streets we aim to stimulate the inside-outside communication from time to time.
This kind of open mechanism not only makes the future street more funny and understandable, but also reserves the public's memory of the site's recent character.

The plan intends to diminish the designer's personal influence, and determines its shape by the design limiting elements of city, economy, culture and technique.

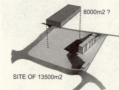

SITE OF 13500m2
8000m2 ?

Standard morphology of bande type building can benefit maximun solar contribution, but it is going to create lots of shadowed area.

普通条式建筑形态能最大化太阳能的利用，可是造成绿化和空地区大量阴影。

BUILT / NON BUILT

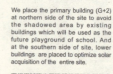

MORPHOLOGY / SHADOW

We place the primary building (G+2) at northern side of the site to avoid the shadowed area by existing buildings which will be used as the future playground of school. And at the southern side of site, lower buildings are placed to optimize solar acquisition of the entire site.

我们尝试将主要建筑部分放在基地北方不受已建建筑遮盖的位置，阴影部分则为为体育运动场所。而南向建筑则采用低层处理，以最大化减少对其后建筑的遮盖。

OPTIMIZATION SOLAR ACQUISITION

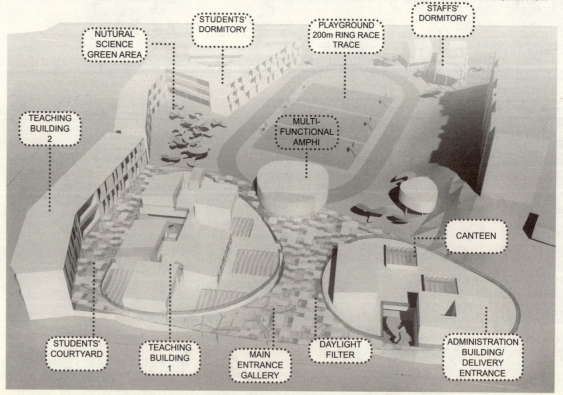

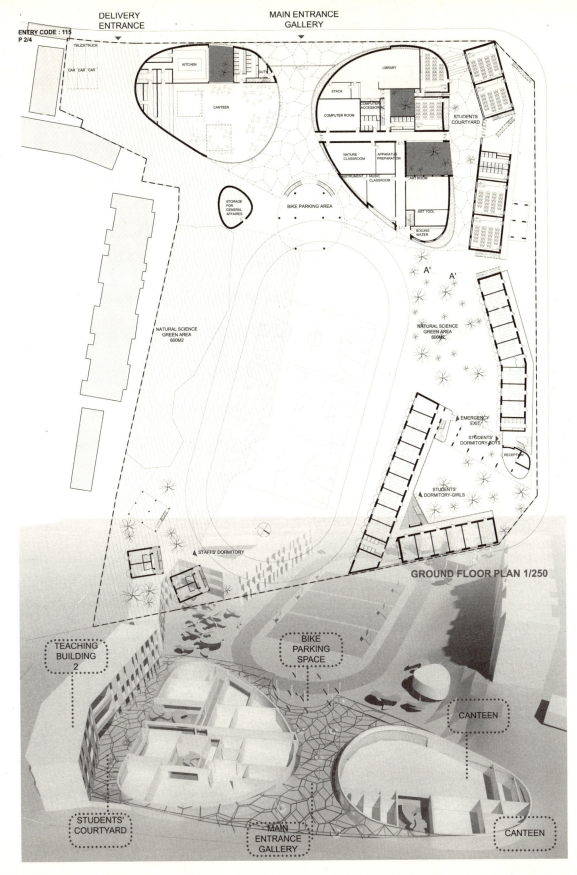

GROUND FLOOR PLAN 1/250

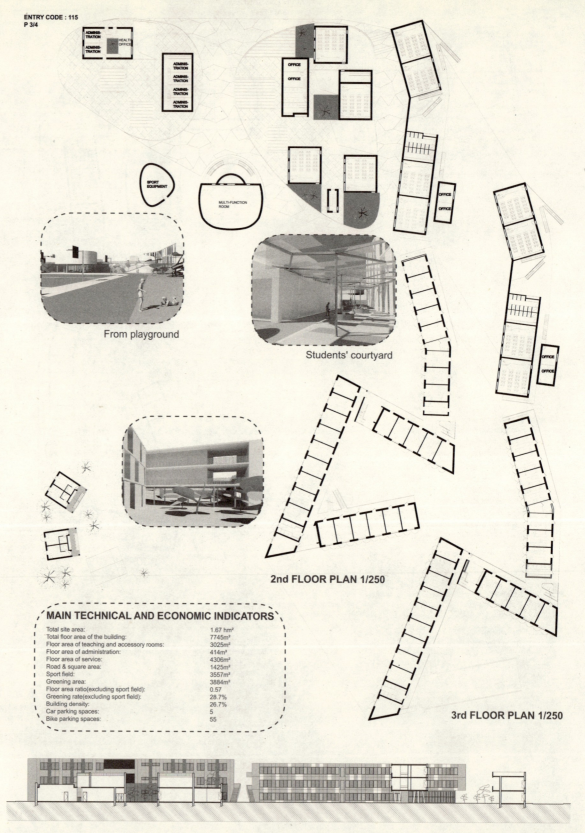

MAIN TECHNICAL AND ECONOMIC INDICATORS

Total site area:	1.67 hm²
Total floor area of the building:	7745m²
Floor area of teaching and accessory rooms:	3025m²
Floor area of administration:	414m²
Floor area of service:	4306m²
Road & square area:	1425m²
Sport field:	3557m²
Greening area:	3884m²
Floor area ratio(excluding sport field):	0.57
Greening rate(excluding sport field):	28.7%
Building density:	26.7%
Car parking spaces:	5
Bike parking spaces:	55

From playground

Students' courtyard

2nd FLOOR PLAN 1/250

3rd FLOOR PLAN 1/250

SECTION PLANE 1/250

NORTHERN ELEVATION 1/250

WESTERN ELEVATION 1/250

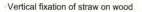

- Vertical fixation of straw on wood
- Louver window in steel cladding
- Steel skin cladding FREQUENCE
- Waterproof layer
- Thermical isulator in straw yellow
- Anti-steam layer
- Finish coat

DETAIL / EXTERIOR WALL 1/25

SCHEMA COMPOSITION

Insulator in yellow straw is a developed construction technique in Europe. 25cm of yellow straw as insulator has the same performance as 20cm of mineral wool.

Straw is a low-price and 100% natural construction material which is available in countryside of Sichuan Province. Recycling yellow straw can not only avoid the farmer from burning straw after harvest to produce more CO_2, but also can provide more job opportunities for local people.

REFERENCES

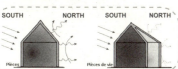

SCHEMA HQE

VIEW / MAIN ENTRANCE GALLERY

优秀奖
Honorable Mention Prize

项目名称：阳光与希望
　　　　　Sunshine and Hope
作　　者：赵增鑫
参赛单位：北京建筑工程学院

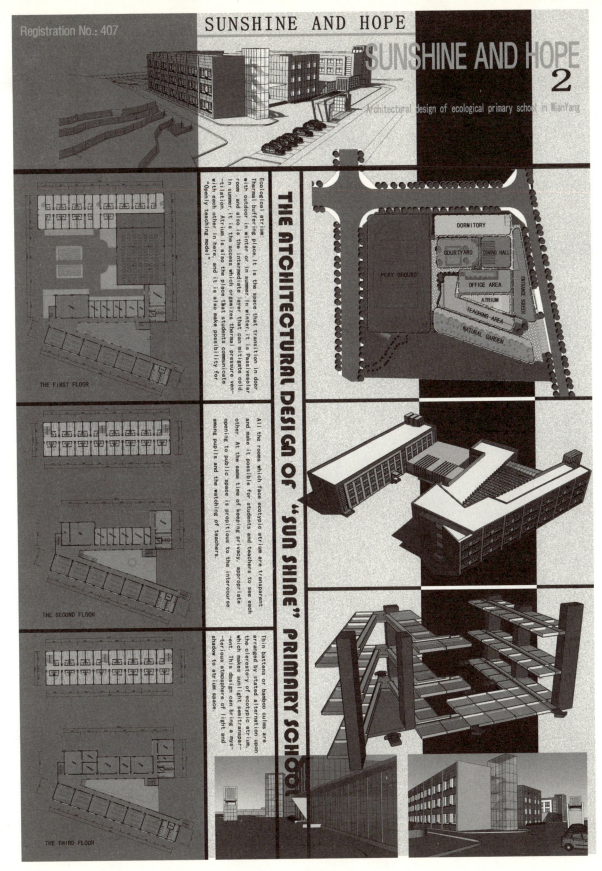

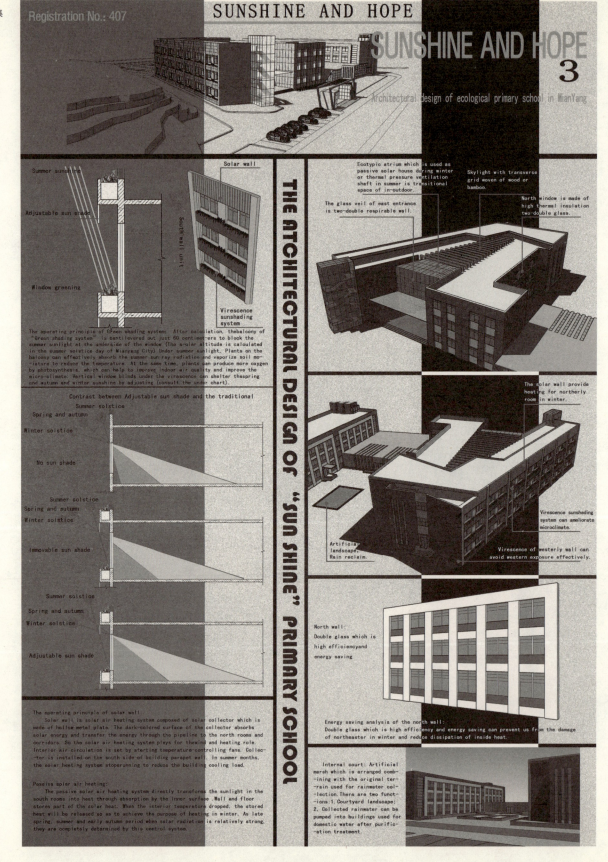

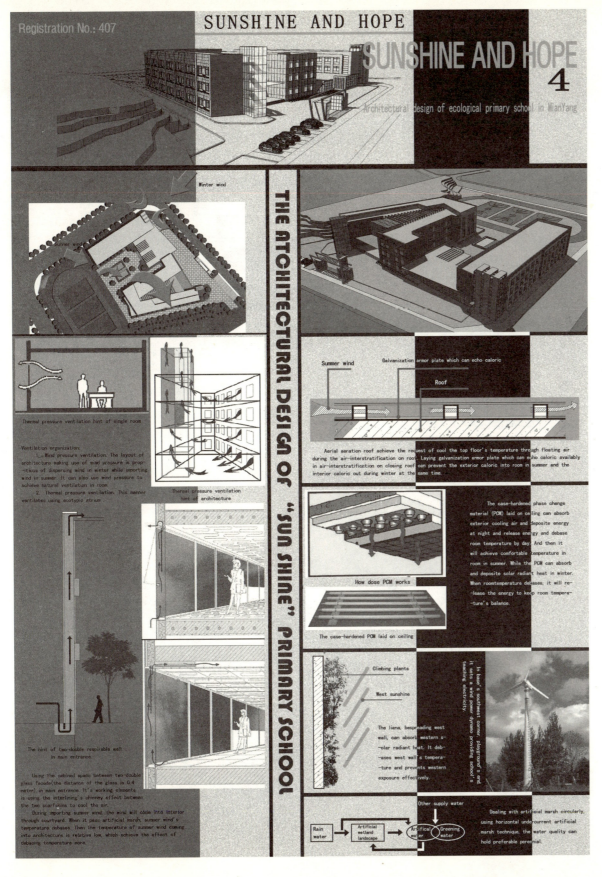

优秀奖
Honorable Mention Prize

项目名称：阳光 & 温暖
　　　　　Light & Warm
作　　者：吴昊、姜成晟、王超、
　　　　　连晓俊、黄荣
参赛单位：上海大学美术学院建筑系

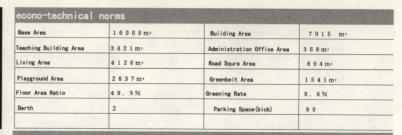

ID:433
Solar School
Light & Warm
Summarization Page
1 2 3 4 5 6

econo-technical norms

Base Area	16000m²	Building Area	7915 m²
Teaching Building Area	3421m²	Administration Office Area	368m²
Living Area	4126m²	Road Squre Area	894m²
Playground Area	2837m²	Greenbelt Area	1541m²
Floor Area Ratio	49.5%	Greening Rate	9.6%
Berth	2	Parking Space(bick)	89

Design notes

本方案为寄宿制学校，由于教学楼和宿舍在太阳能的使用方式不尽相同，在设计策略上也得出不同的结果。教学楼部分尽量使采光和通风做到最大化，我们采用了南北光隔转构件来保持室内照度的均匀，而"呼吸走廊"能让热气流上升，形成气流循环，结合棱风中庭加大通风效果，走廊适当加宽，是学生理想的活动场所。宿舍部分白天基本"待机"，怎样储存太阳能是我们考虑的一个重点，所以我们将一些已经成熟的技术合理的运用其中，如特朗布墙、蓄热墙、蓄热屋顶等。

As this is a boarding school, the solar energy usage mode of teaching building is different from dormitory, so the design strategy is different. We maximize the availability factor of lighting and ventilation in teaching building design, and use light deflecting elements to keep illumination uniformity indoor. Moreover "breathing corridor" can make hot airflow rise to form cycle draught, combine with the atrium to strengthen ventilation. Widen the corridor to let students own a ideal stage. Basically the part of dormitory is "readiness", it is a emphasis how to store the solar energy, so we use some muture technology in the design, such as trombe wall, heat preservation wall, heat preservation roof and so on.

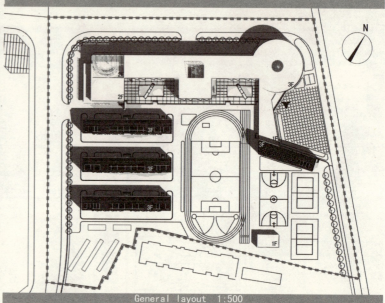

General layout 1:500

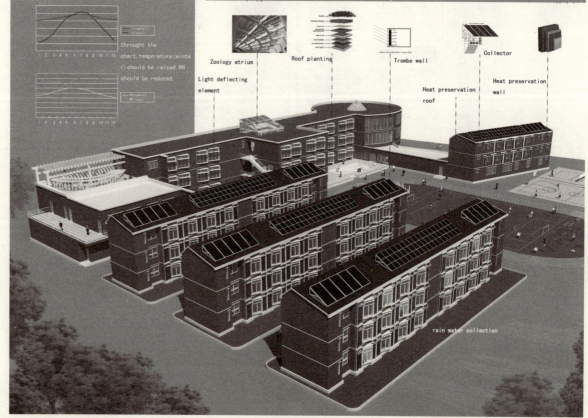

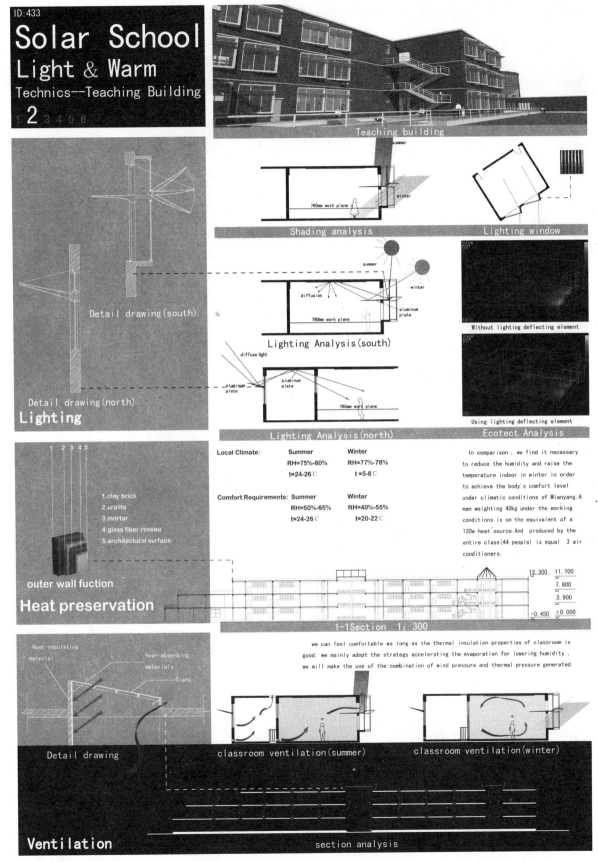

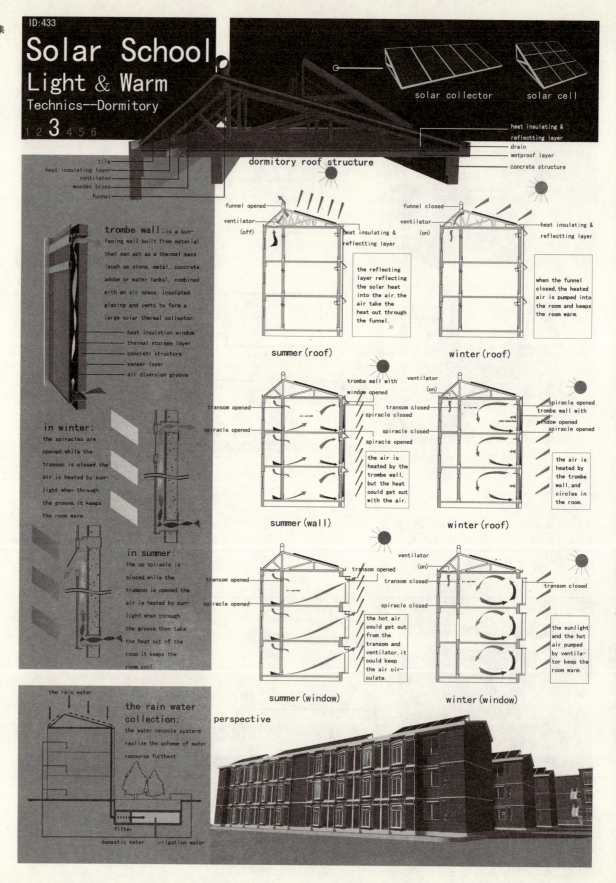

Solar School
Light & Warm
Architecture Page 1
1 2 3 **4** 5 6

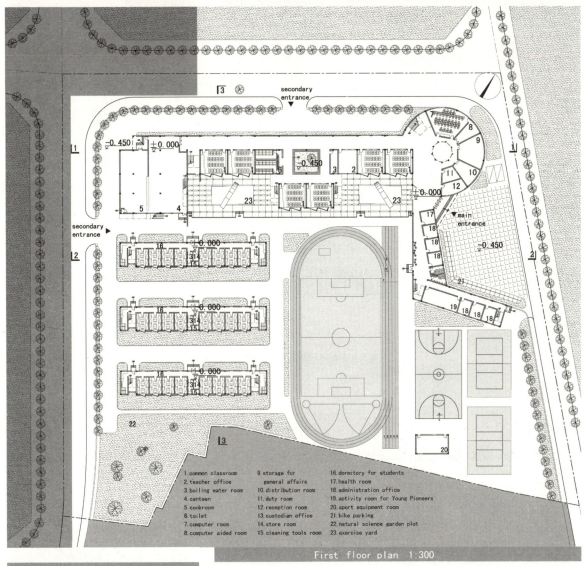

1. common classroom
2. teacher office
3. boiling water room
4. canteen
5. cookroom
6. toilet
7. computer room
8. computer aided room
9. storage for general affairs
10. distribution room
11. duty room
12. reception room
13. custodian office
14. store room
15. cleaning tools room
16. dormitory for students
17. health room
18. administration office
19. activity room for Young Pioneers
20. sport equipment room
21. bike parking
22. natural science garden plot
23. exercise yard

First floor plan 1:300

We creat two atriums in the part of special teaching building and the part of common classroom. They are the center of the coverway. First, they make the space changeful, and form a space for students to settle. Second, they put more activities into the boring space. We hope that the two atriums can form a absorbing locale together with the north coverway. Because, it is not only the come-and-go space, but also a locale full of memory for students.

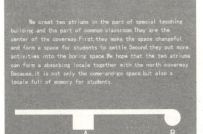

perspective of classroom building atrium(A)

perspective of the entrance atrium(B)

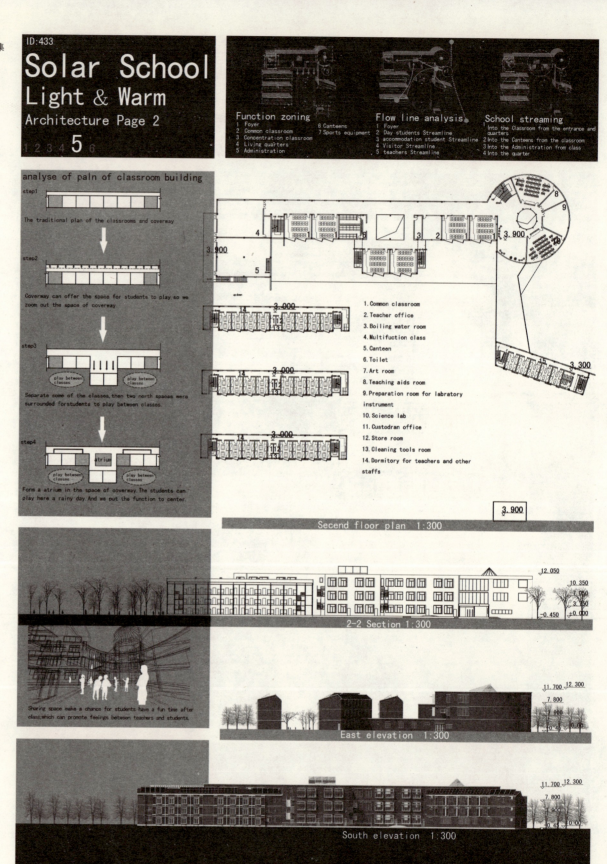

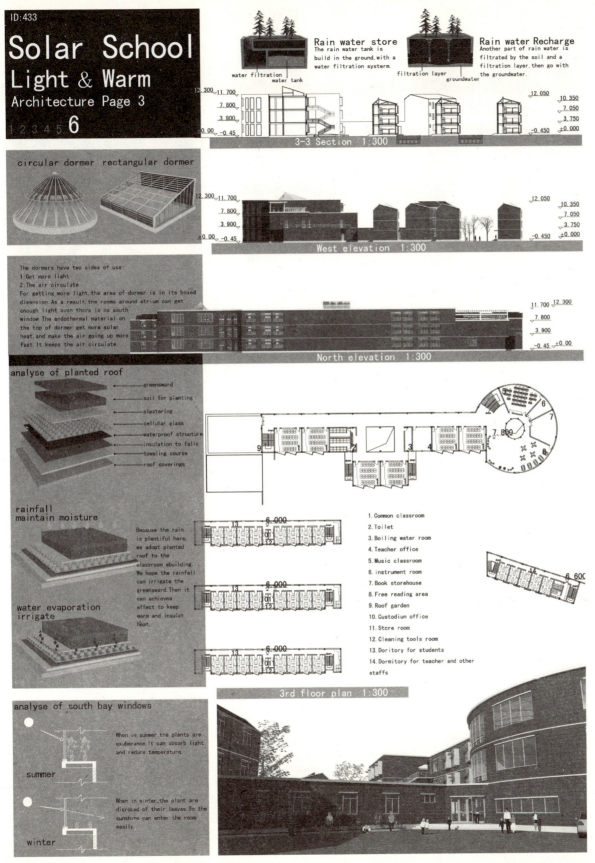

优秀奖
Honorable Mention Prize

项目名称：绿野
　　　　　Green Field
作　　者：万博、刘菁
参赛单位：重庆大学建筑城规学院

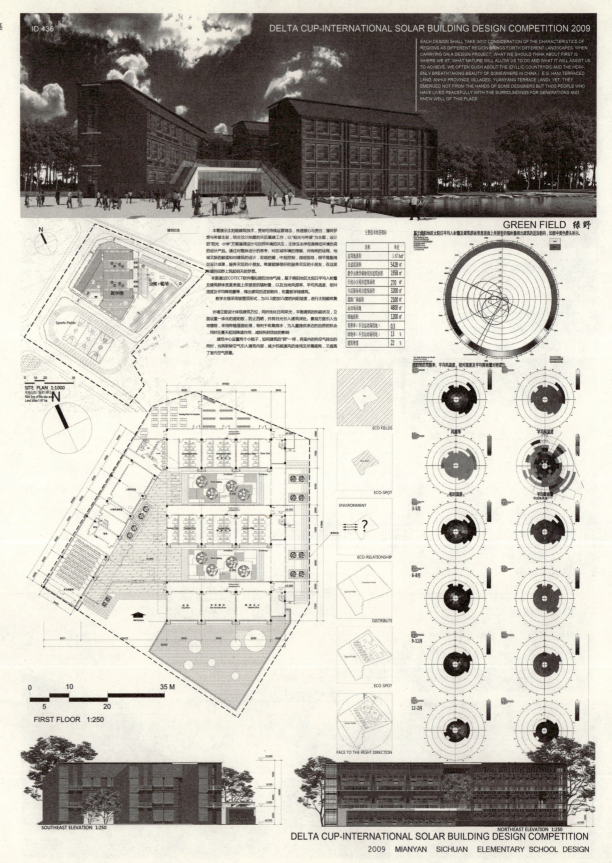

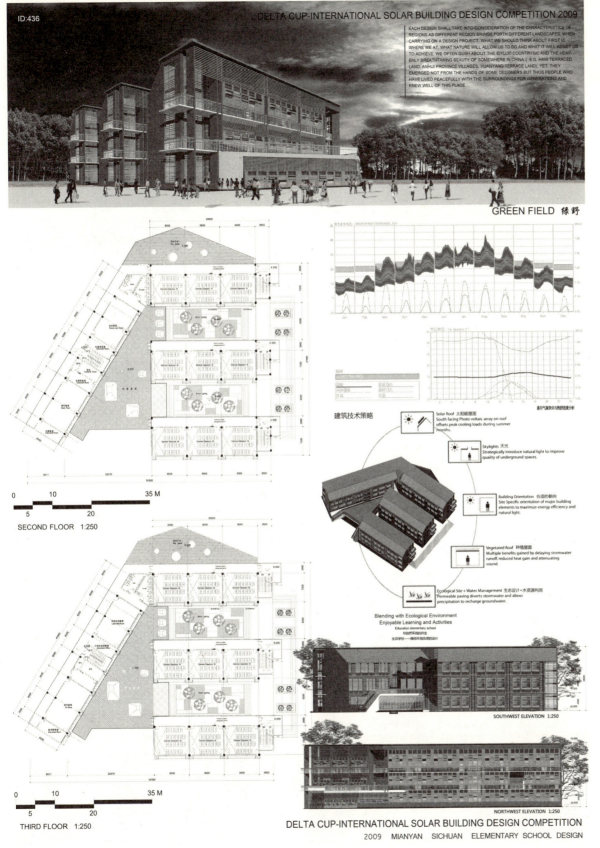

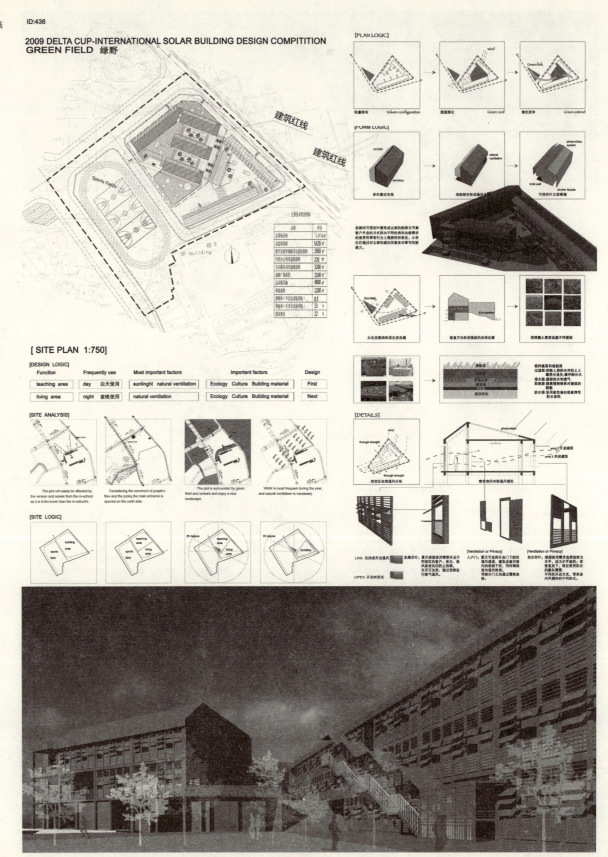

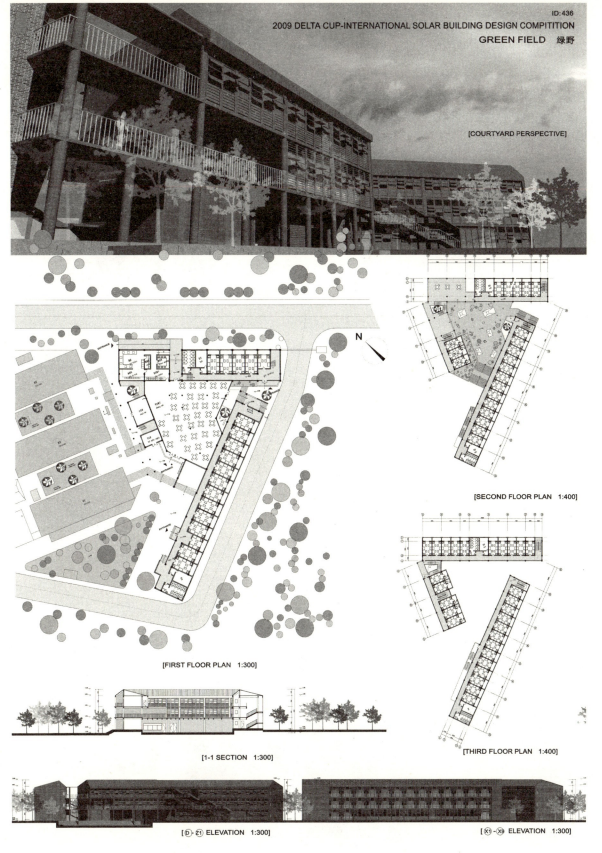

2009 Delta Cup_Solar Building Design Competition

优秀奖
Honorable Mention Prize

项目名称：553号作品
No.553

作　　者：Nicola Bettini、Clara Masotti、
Stefano Brunell、Antonio Bandini、
Stefano Massa、Debora Venturi

参赛单位：Architect n°3230 - Ordine degli
Architetti di Bologna (IT)

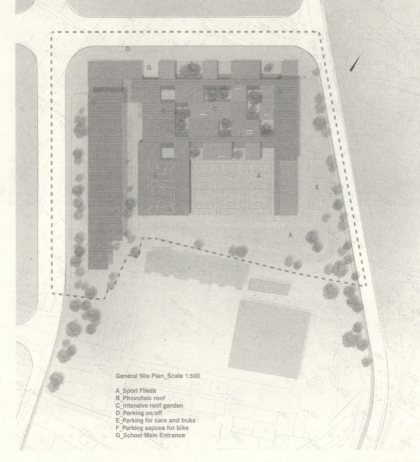

General Site Plan_Scale 1:500
A_Sport Fileds
B_Phovoltaic roof
C_Intensive roof garden
D_Parking on/off
E_Parking for cars and truks
F_Parking sapces for bike
G_School Main Entrance

Mianyang_Site analysis: Lat 31.5° Long 104.7°

Site Climate analysis:

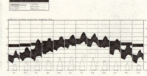
Monthly Diurnal Averages Climate Data

Direct Solar Radiation Diagramme

Relative Humidity Diagramme

BRIEF

General criteria
- Modularity: simplicity and cheapness of design, rapidity of construction, and effectiveness of facilities are the driving forces of this project.
- Functionality: the building accomplishes its educational goals according to a rational and at the same time flexible approach.
- Environmental quality: the aim of this project is to design a building that may merge with the surrounding natural landscape with best answer to hearthquake sollecitation, while meeting the highest energy efficiency standards.

The project:
The new school is conceived as a system of "green" elements that are organized in more levels.
The project's matrix is based on an articulated structure of classrooms and patios which allows equilibration between the relevant covered and open spaces.
The courtyards and the green roof are conceived as proper gardens where educational activities may be further developed in an open air environment.

The Energetic approach:
The form, orientation, constructional and distributive features of the building are defined on the basis of a thorough analysis of the site and its climatic and environmental factors.
Basically, the building develops on a single floor covered with a garden roof that allows for high performance thermal and acoustic insulation, and it is articulated in a system of internal courtyards that provide the whole construction with optimal illumination and ventilation.
The highest energy efficiency standards are achieved through special technologies and installations that combine passive functioning modes with innovative technological systems.

Stereographic Diagramme : Average Temperature January - June

Optimum Orientation based on average daily incident radiation on a vertical surface

Overview of the Green roof of the Solar School

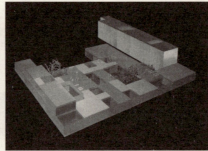
Overview of the Solar School

Stereographic Diagramme : Average Temperature Juny - December

Prevailing Winds Frequency (Hrs)

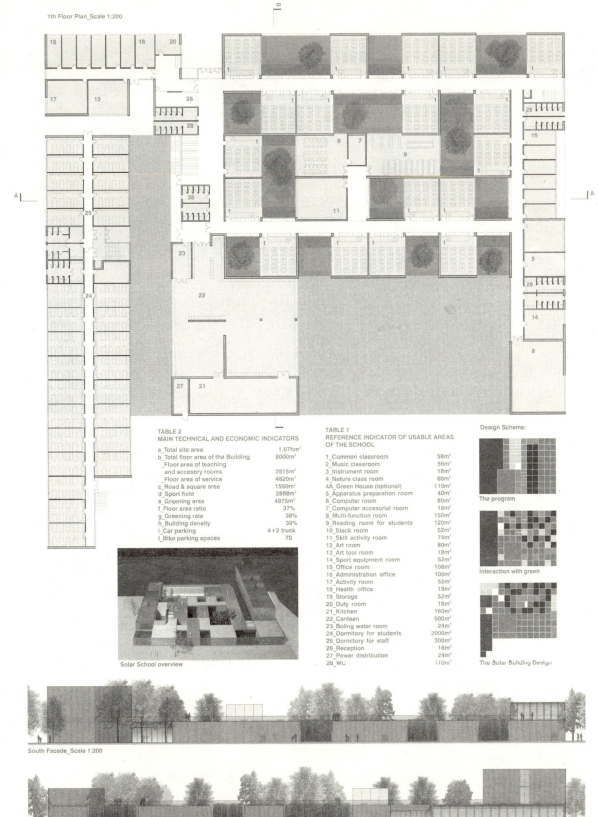

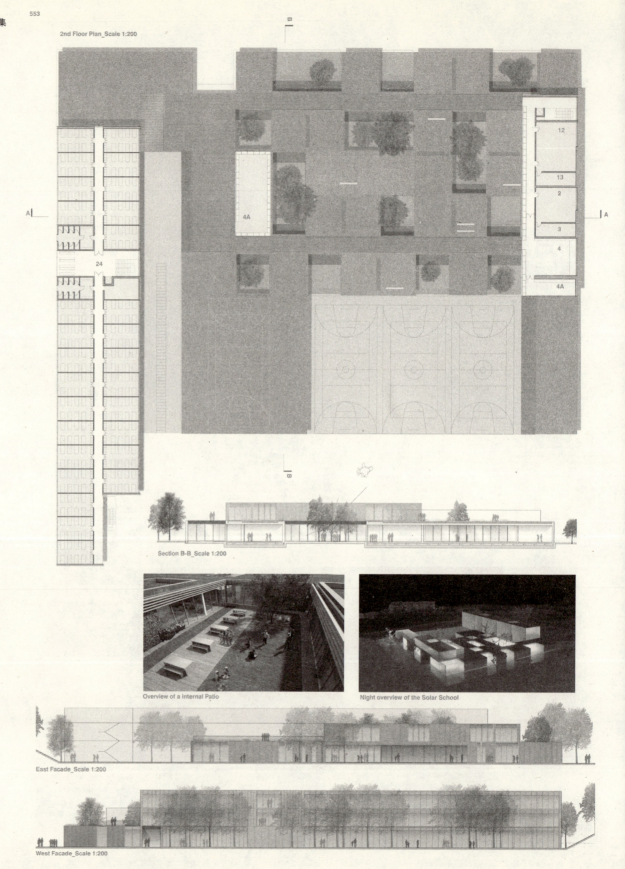

3th Floor Plan_Scale 1:200

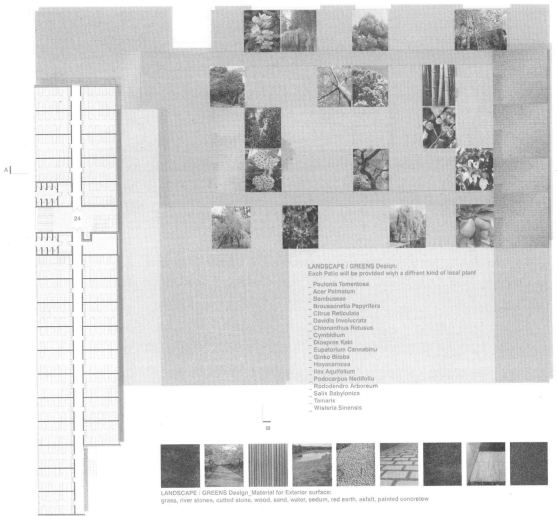

LANDSCAPE / GREENS Design:
Each Patio will be provided wiyh a diffrent kind of local plant

_ Paulonia Tomentosa
_ Acer Palmatum
_ Bambuseae
_ Broussonetia Papyrifera
_ Citrus Reticulata
_ Davidia Involucrata
_ Chionanthus Retusus
_ Cymbidium
_ Diospros Kaki
_ Eupatorium Cannabinu
_ Ginko Biloba
_ Hoyacarnosa
_ Ilex Aquifolium
_ Podocarpus Neriifoliu
_ Rododendro Arboreum
_ Salix Babylonica
_ Tamarix
_ Wisteria Sinensis

LANDSCAPE / GREENS Design_Material for Exterior surface:
grass, river stones, cutted stone, wood, sand, water, sedum, red earth, asfalt, painted concretew

View of a Patio of the Solar School

View of a Patio throught the Classroom

Section A-A_Scale 1:200

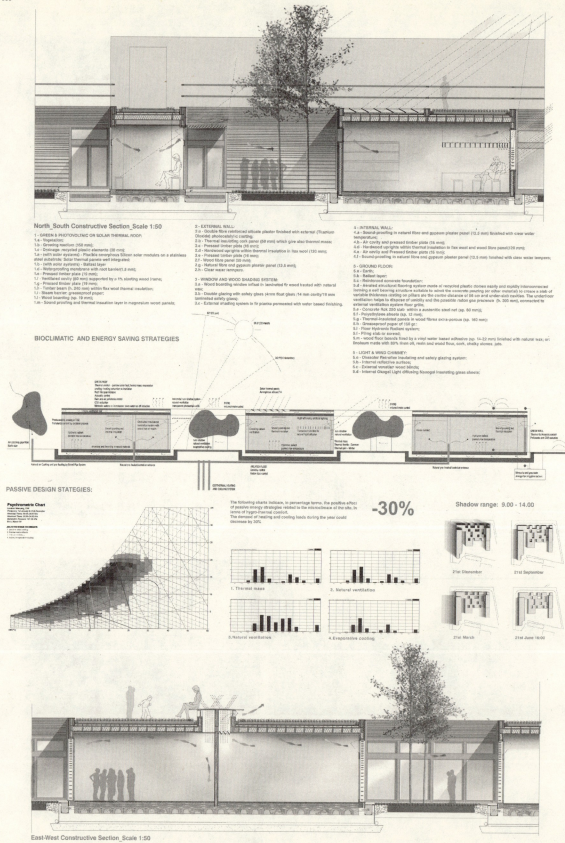

PASSIVE AND ACTIVE PLANTS INTEGRATION SCHEME:

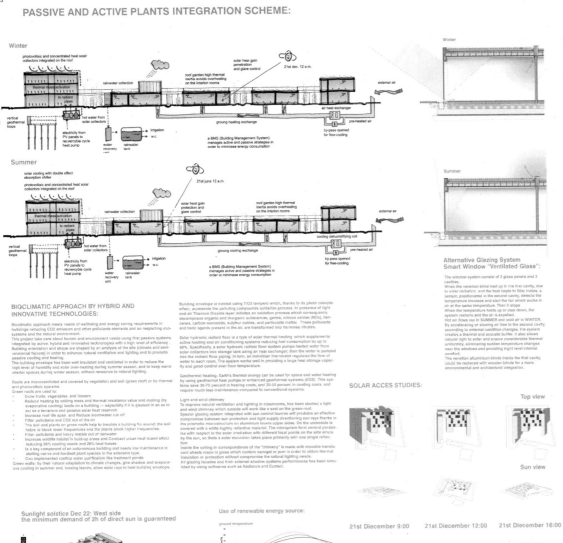

BIOCLIMATIC APPROACH BY HYBRID AND INNOVATIVE TECHNOLOGIES:

Bioclimatic approach meets needs of wellbeing and energy saving requirements in buildings reducing CO2 emission and other pollutants elements and so respecting ecosystems and the natural environment.
This project take care about human and environment needs using first passive systems integrated by active, hybrid and innovative technologies with a high level of efficiency.
Building orientation and shape are shaped by an analysis of the site (climate and environmental factors) in order to enhance natural ventilation and lighting and to promote passive cooling and heating.
The building envelope has been well insulated and ventilated in order to reduce the high level of humidity and solar over-heating during summer season, and to keep warm interior spaces during winter season, without renounce to natural lighting.

Roofs are microventilated and covered by vegetation and soil (green roof) or by thermal and photovoltaic systems.
Green roofs are used to:
- Grow fruits, vegetables, and flowers
- Reduce heating by adding mass and thermal resistance value and cooling (by evaporative cooling) loads on a building — especially if it is glassed in so as to act as a terrarium and passive solar heat reservoir
- Increase roof life span and Reduce stormwater run off.
- Filter pollutants and CO2 out of the air
- The soil and plants on green roofs help to insulate a building for sound; the soil helps to block lower frequencies and the plants block higher frequencies.
- Filter pollutants and heavy metals out of rainwater
- Increase wildlife habitat in built-up areas and Combact urban heat island effect reducing 26% cooling needs and 26% heat losses
- Is a key component of an autonomous building and needs low maintenance in stuffing native and hardiest plant species in the extensive type.
- Can implemented rooftop water purification like treatment ponds

Green walls: by their natural adaptation to climate changes, give shadow and evaporative cooling in summer and, loosing leaves, allow solar rays to heat building envelope.

Building envelope is treated using TiO2 tempera which, thanks to its photo catalytic effect, accelerate the polluting compounds oxidation process. In presence of light and air Titanium Dioxide layer initiates an oxidation process which consequently decomposes organic and inorganic substances, germs, nitrous oxides (NOx), benzenes, carbon monoxide, sulphur oxides, and particulate matter. These pollutants and toxic agents present in the air, are transformed into harmless nitrates.

Solar hydronic radiant floor is a type of solar thermal heating which supplements active heating and air conditioning systems reducing fuel consumption by up to 90%. Specifically, a solar hydronic radiant floor system pumps heated water from solar collectors into storage tank using an heat exchanger; then the water is pumped into the radiant floor piping. In all, an individual thermostat regulates the flow of water to each room. The system works well in providing a large heat storage capacity and good control over floor temperature.

Geothermal heating: Earth's thermal energy can be used for space and water heating by using geothermal heat pumps or enhanced geothermal systems (EGS). This systems save 30-70 percent in heating costs, and 20-50 percent in cooling costs, and require much less maintenance compared to conventional systems.

Light and wind chimney
To improve natural ventilation and lighting in classrooms, has been studied a light and wind chimney which outside will work like a seat on the green-roof.
Special glazing system integrated with sun control louvres will provides an effective compromise between sun protection and light supply directioning sun rays thanks to the prismatic microstructure on aluminium louvre upper sides. On the underside is covered with a white highly reflective material. The microprism form several parabolas with respect to the solar irradiation with different focal points on the site struck by the sun, so thats a solar excursion takes place primarily with one single reflection
Inside the ceiling in correspondence of the "chimeny" is made with movable translucent sheets made in glass which contains nanogel or pcm in order to obtain thermal insulation or protection without compromise the natural lighting needs.
All glazing facades and their external shadow systems performances has been simulated by using softwares such as Radiance and Ecotect.

Alternative Glazing System
Smart Window "Ventilated Glass":

The window system consist of 3 glass panels and 2 cavities.
When the venetian blind heat up in the first cavity, due to solar radiation and the heat begin to filter inside, a sensor, positioned in the second cavity, detects the temperature increase and start the fan which sucks in air at the same temperature. Then it stops.
When the temperature heats up or cool down, the system restarts and the air is expelled.
Hot air flows out in SUMMER and cold air in WINTER.
By accelerating or slowing air flow in the second cavity, according to external condition changes, the system creates a thermal and acoustic buffer. It also allows natural light to enter and ensure considerable thermal uniformity, eliminating sudden temperature changes near the windows and providing bright level internal comfort.
The venetian alluminium blinds inside the first cavity, could be replaced with wooden blinds for a more environmental and architectural integration.

SOLAR ACCES STUDIES:

Sunlight solstice Dec 22: West side
the minimum demand of 2h of direct sun is guaranteed

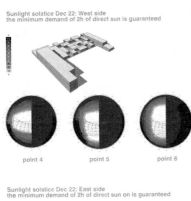

Use of renewable energy source:

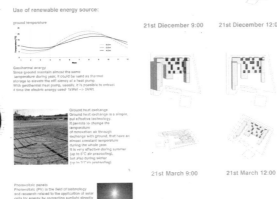

Geothermal energy
Since ground maintain almost the same temperature during year, it could be used as thermal storage to elevate the efficiency of a heat pump.
With geothermal heat pump, usually, it is possible to extract 4 time the electric energy used: 1aWel --> 5kWt.

Ground heat exchange
Ground heat exchange is a simple, but effective technology.
It permits to change the temperature of renovation air through exchange with ground, that have an almost constant temperature during the whole year.
It is very effective during summer (up to 5°C air precooling), but also during winter (up to 7-7°C preheating).

Photovoltaic panels
Photovoltaic (PV) is the field of technology and research related to the application of solar cells for energy by converting sunlight directly into electricity.
Despite of the relatively low efficiency, with the use of photovoltaic generated energy is possible to locally produce electricity without any additional cost for fuel and without CO2 atmosphere emissions.
Today in the market it can be found different type and shape of the panel, making them usable on different position in buildings.

Solar thermal collectors
Solar thermal collector is a technology for harnessing solar energy for thermal energy.
It is typically used for domestic hot water production, but it could be applied in conjunction with LTDS radiant panel, low temperature AHU, but it's very useful for desiccant chemical wheel regeneration. Solar thermal collector allow to produce low cost energy without traditional fuel and CO2 emissions.

Sunlight solstice Dec 22: East side
the minimum demand of 2h of direct sun is guaranteed

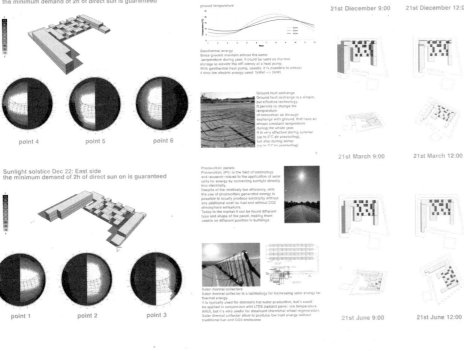

优秀奖
Honorable Mention Prize

项目名称：绵阳阳光小学
　　　　　The Sunshine Primary School in Mianyang
作　者：龚浩朋、龚浩斌
参赛单位：中联设计顾问有限公司

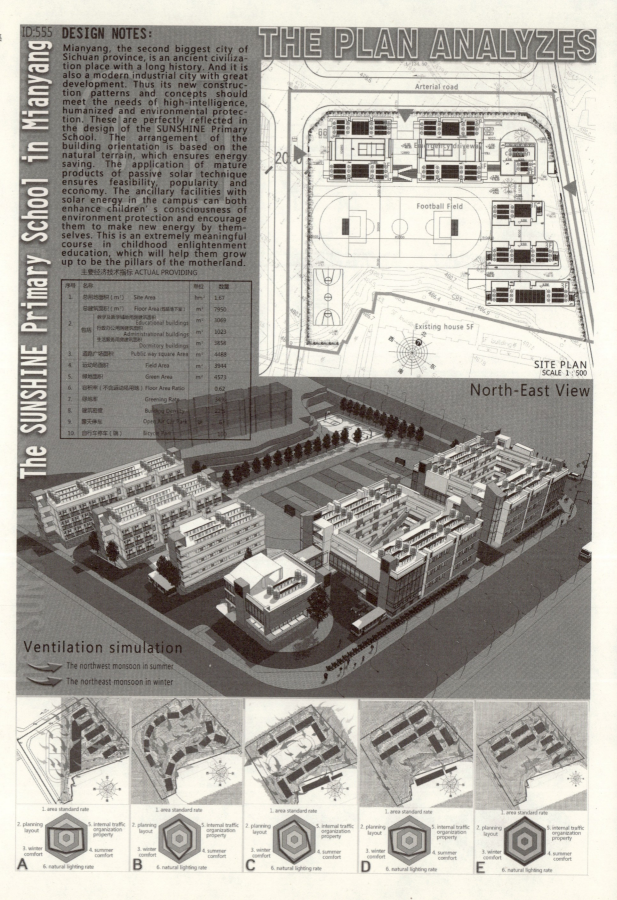

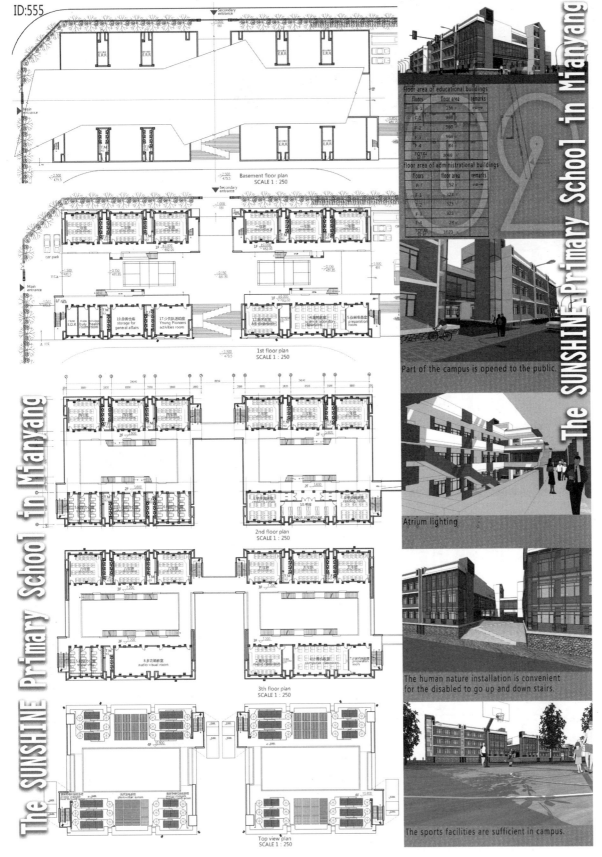

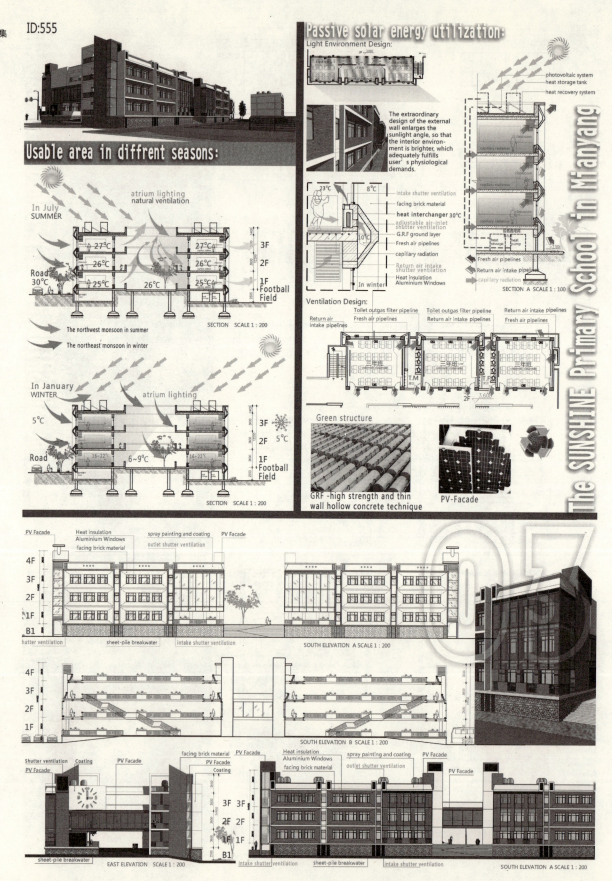

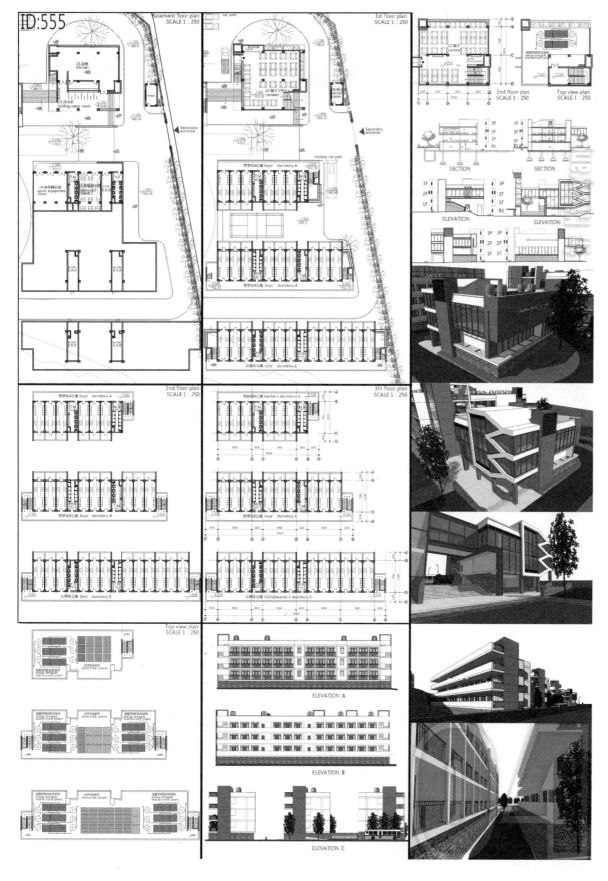

优秀奖
Honorable Mention Prize

项目名称：光织品
　　　　　Light Fabric
作　者：曾雪松、段希莹
参赛单位：重庆大学建筑城规学院

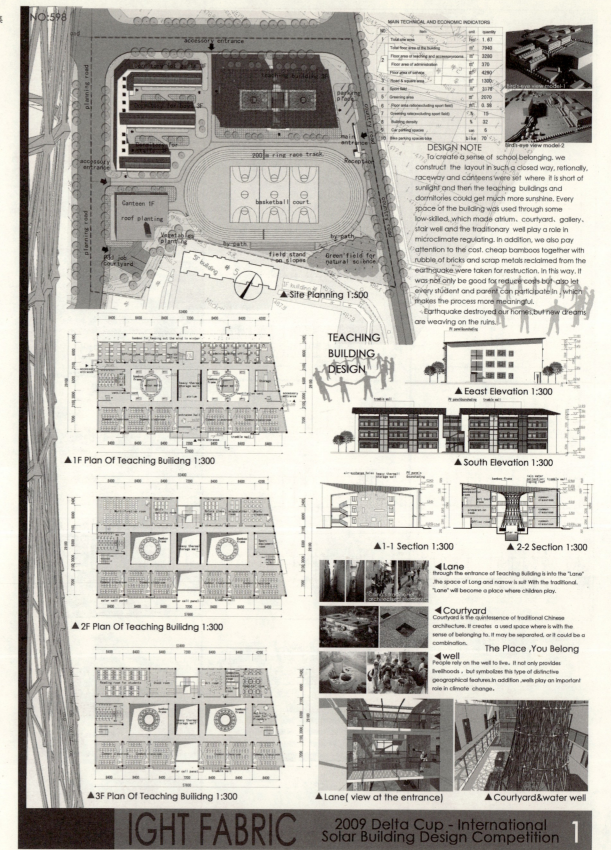

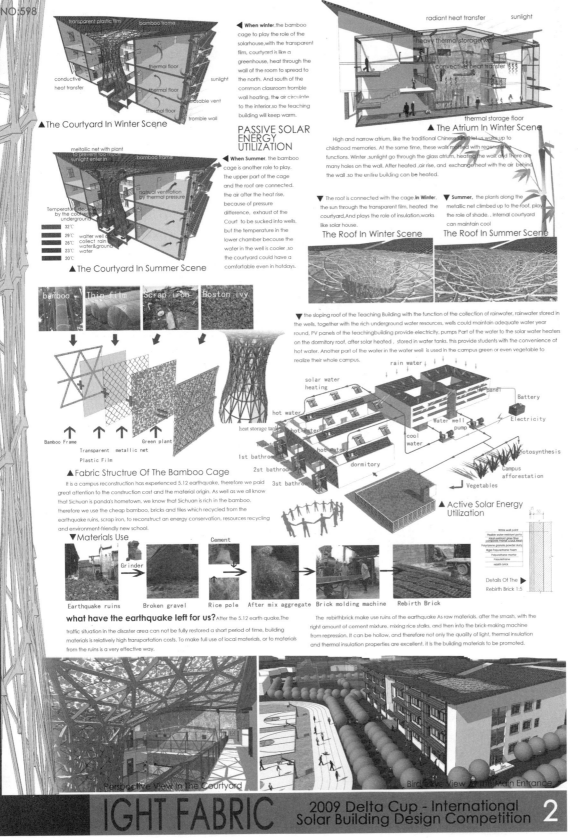

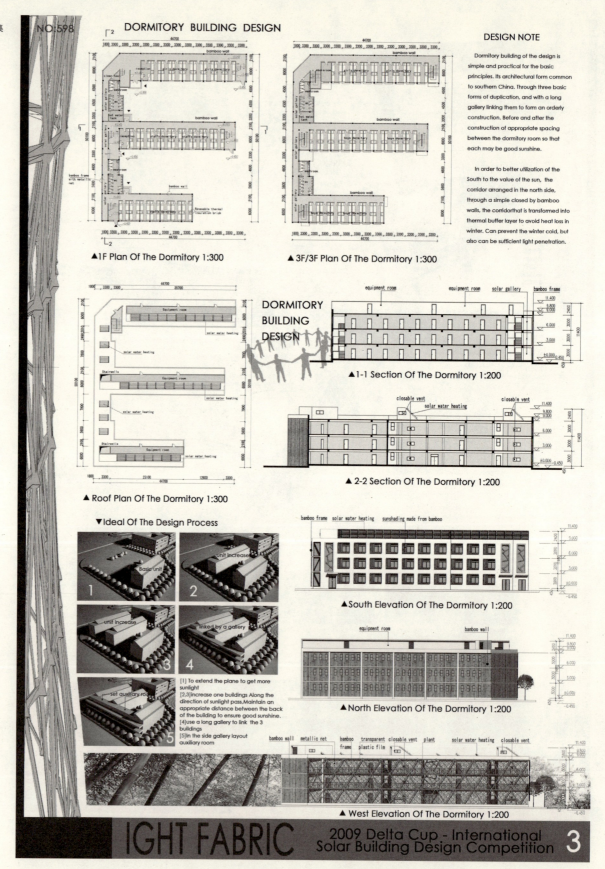

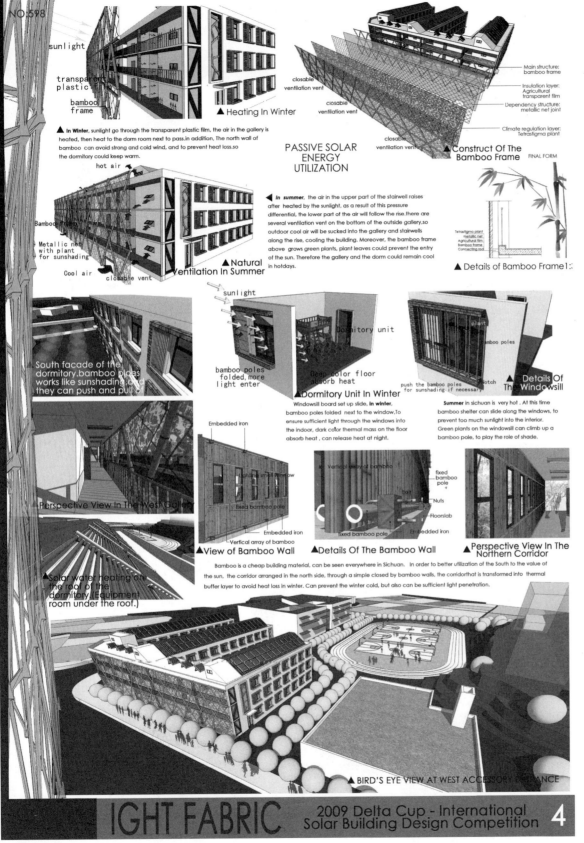

优秀奖
Honorable Mention Prize

项目名称：阳光小学
　　　　　Sunshine Primary School
作　者：包藏新、钟光浒
参赛单位：东南大学建筑学院

设计说明

这是一个很特殊的环境：一方面绵阳处于我国太阳能资源最稀缺的四川盆地，相对于主动式太阳能利用，被动式太阳能利用更占主导地位。在建筑形体上强化自然通风；通过遮阳、建筑外围护结构保温等方式最大限度地实现建筑的保温隔热，从而达到节能的目的。主动式太阳能主要用于宿舍集热以及太阳能路灯。另一方面，大量利用当地盛产的竹，与当地的文化传统产生呼应，同时在一定程度上促进了当地传统建造的发展。

Design Specification
It is a special project. On one hand, Mianyang is located in the Sichuan Basin, where the solar energy is the lowest in the country. Compared with active solar energy utilization, passive solar energy utilization is more dominant. We adjust our building in order to increase the natural ventilation. We adopt shading, thermal insulation such as LOW-E window on the building enclosure to save energy. Active acquired solar energy is mostly used in students' dormitory and road lights.
  On the other hand, we use a lot of bamboo in our building. Not only does it respond to the local culture, but it more or less promotes the development of local traditional construction.

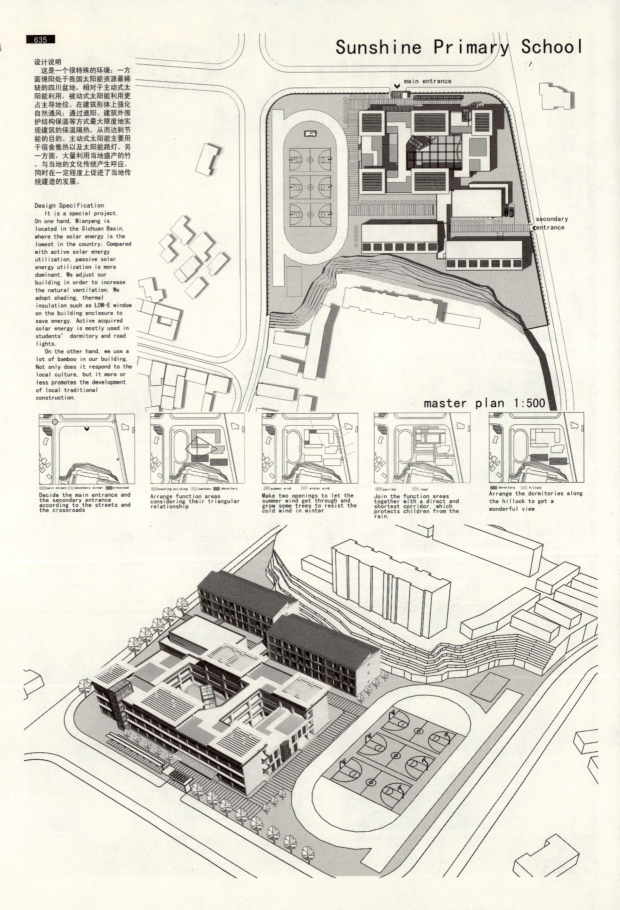

Sunshine Primary School

master plan 1:500

Decide the main entrance and the secondary entrance according to the streets and the crossroads.

Arrange function areas considering their triangular relationship.

Make two openings to let the summer wind get through and grow some trees to resist the cold wind in winter.

Join the function areas together with a direct and shortest corridor, which protects children from the rain.

Arrange the dormitories along the hillock to get a wonderful view.

1 view at the entrance on campus.

2 view in the teaching building.

3 view from the northeast of the teaching building.

4 view between the teaching building and the sport yard.

5 view between the teaching building and the dormitory.

6 view in the corridor from the teaching building.

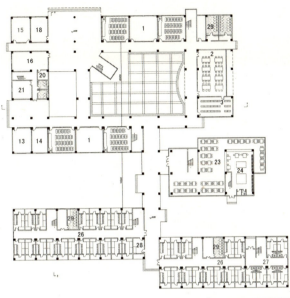

1st floor plan 1:300

1 common classroom 2 reading room for students 3 book storehouse
4 music classroom 5 instrument room 6 laboratory 7 preparation room
8 labor skill room 9 arts classroom 10 computer classroom
11 multi-function room 12 teachers' office 13 radio room
14 sport equipment room

15 reception room 16 administration entrance 17 administration office
18 activity room for Young Pioneers 19 health room 20 duty room
21 storage for general affairs 22 management

23 canteen 24 kitchen 25 boiling water room 26 students' dormitory
27 hostel for single staffs 28 reception room for dormitory 29 WC

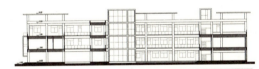
section a-a 1:300

section b-b 1:300

south elevaton 1:300

north elevaton 1:300

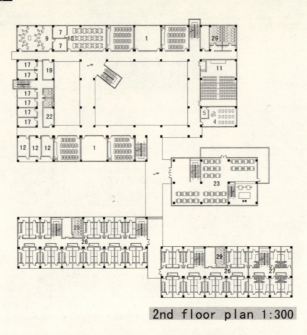

2nd floor plan 1:300

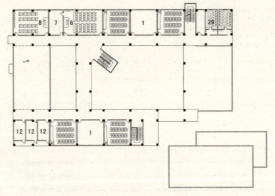

3rd floor plan 1:300

east elevaton 1:300

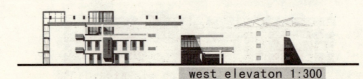

west elevaton 1:300

Distribution of solar energy in China

From this picture we can learn that the solar energy efficiency in Sichuan is the lowest in China, thus the photovoltaic technology is not an economical way to save energy. Solar energy heating and solar-powered light are more practical.

Solar energy heating

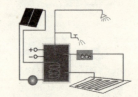

evacuated tube solar collector / flat plate solar energy collector

considering the budget and the climate of Mianyang, we decide to choose the flat plate solar energy collecting system.

the solar heat collecting panels on the roof of the dormitory.

Solar-powered light

The teaching building

Building enclosure system

according to certain specifications, we calculated the thickness of EPS insulating panel.

Roof planting and rainwater recycling

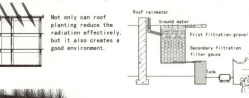

Not only can roof planting reduce the radiation effectively, but it also creates a good environment.

The students' dormitory

Ventilation

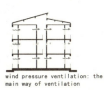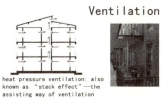

wind pressure ventilation: the main way of ventilation

heat pressure ventilation: also known as "stack effect" — the assisting way of ventilation

Roof construction

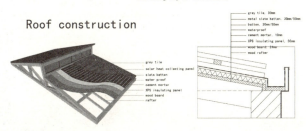

Bamboo & Steel

Bamboo is a kind of material widely used in traditional Chinese buildings, while steel is quite modern. what can we make by joining them together?

1. rail

2. horizontal blind

3. vertical shutter

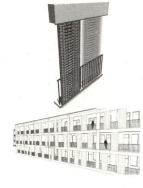

techno-economic indicator

1	site area	1.67hm²
2	total building area	7983m²
	teaching and assisting room area	4329m²
	administration room area	423m²
	living and service area	3231m²
3	roads and squares area	949m²
4	schoolyard area	3268m²
5	planting area	8959.3m²
6	floor area ratio	0.47
7	greening rate	53.6%
8	building density	21.1%
9	car parking amount	6
10	bike parking amount	50

优秀奖
Honorable Mention Prize

项目名称：童年的梦想
The Dream of the Childhood
作　　者：李鹏、郑彬、孙雪梅
参赛单位：东南大学建筑学院

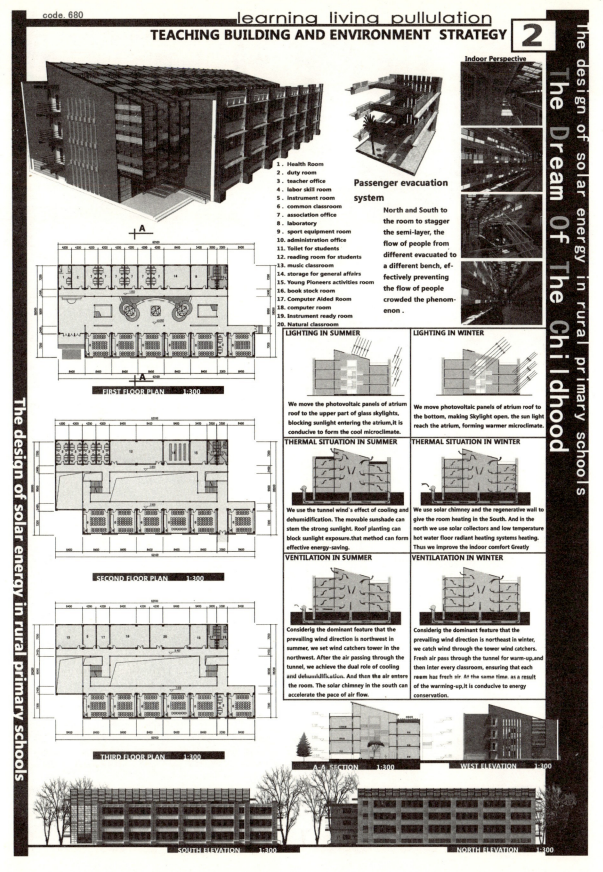

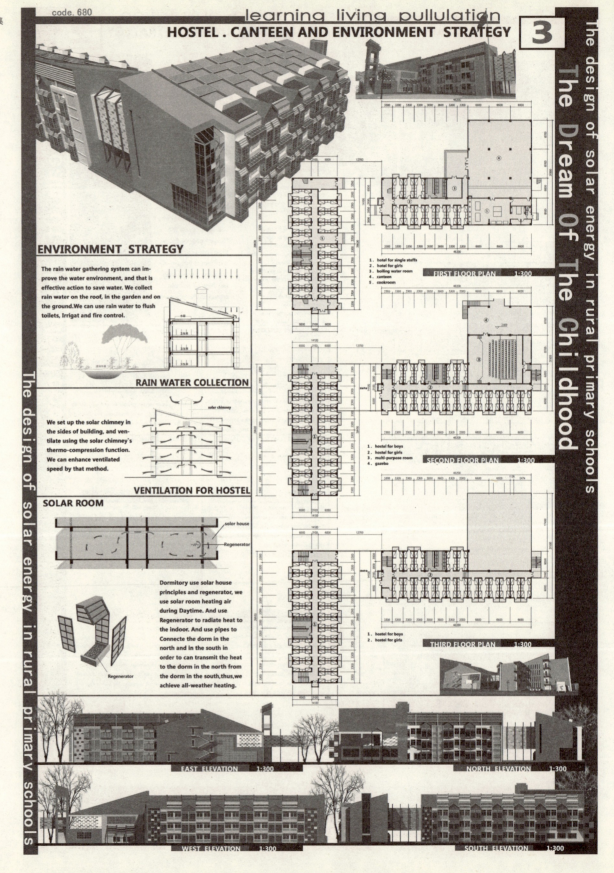

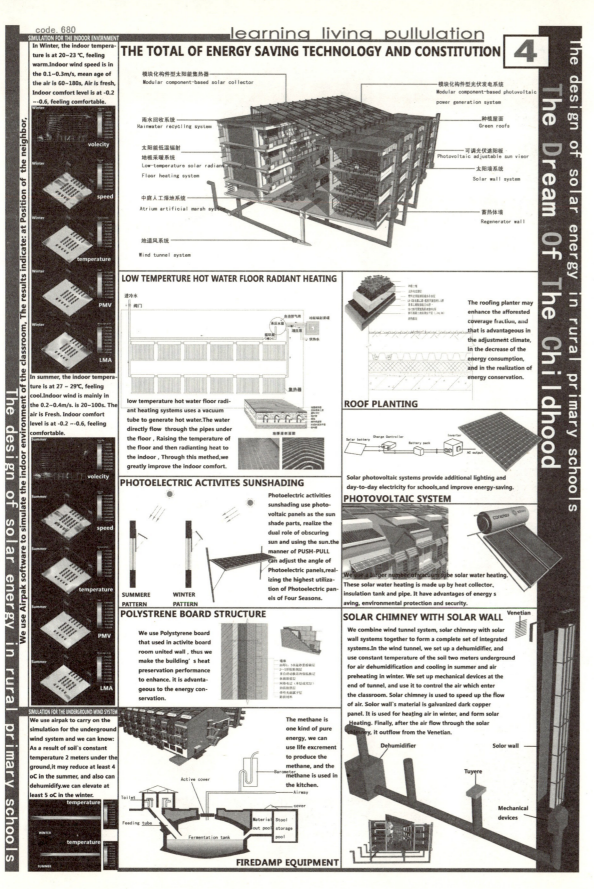

优秀奖
Honorable Mention Prize

项目名称：应变建筑
　　　　　Climate-responsive building
作　者：丁瑜、徐斌、崔陇鹏、连小鑫
参赛单位：东南大学建筑学院

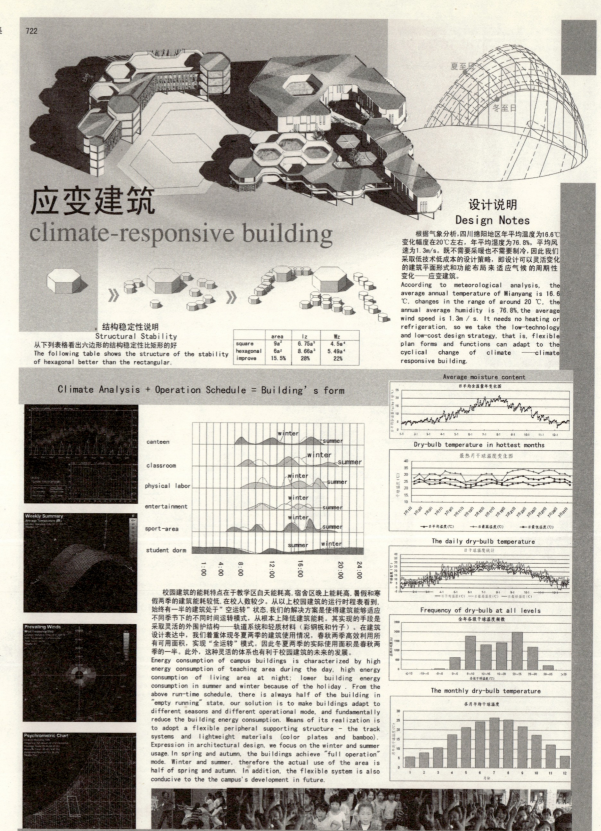

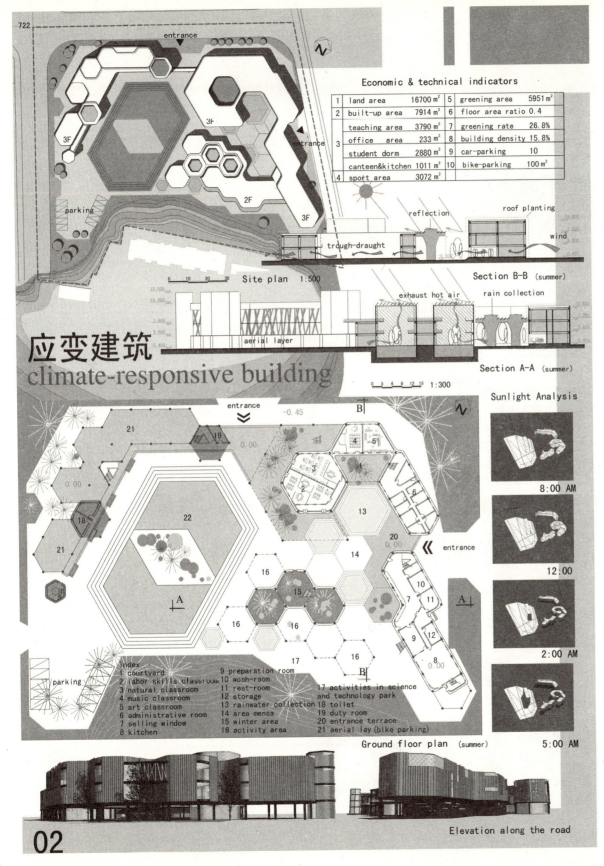

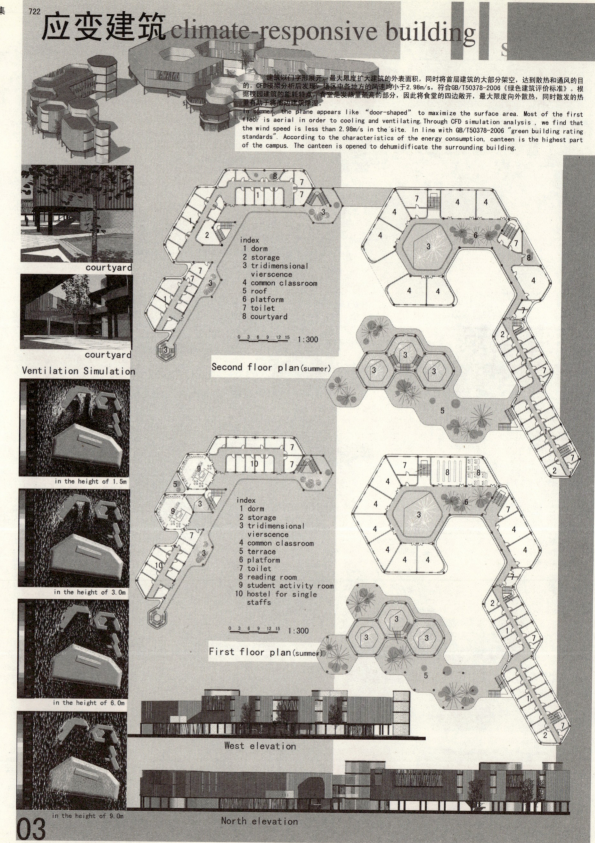

应变建筑 climate-responsive building

建筑以"门"字形展开，最大限度扩大建筑的外表面积，同时将首层建筑的大部分架空，达到散热和通风的目的。CFD模拟分析后发现，场区中各地方的风速均小于2.98m/s，符合GB/T50378-2006《绿色建筑评价标准》。根据校园建筑的能耗特点，食堂是发热量最高的部分，因此将食堂的四边敞开，最大限度向外散热，同时散发的热量有助于将周边建筑降温。

In summer, the plane appears like "door-shaped" to maximize the surface area. Most of the first floor is aerial in order to cooling and ventilating. Through CFD simulation analysis, we find that the wind speed is less than 2.98m/s in the site. In line with GB/T50378-2006 "green building rating standards". According to the characteristics of the energy consumption, canteen is the highest part of the campus. The canteen is opened to dehumidificate the surrounding building.

courtyard

courtyard

Ventilation Simulation

in the height of 1.5m

in the height of 3.0m

in the height of 6.0m

in the height of 9.0m

index
1 dorm
2 storage
3 tridimensional vierscence
4 common classroom
5 roof
6 platform
7 toilet
8 courtyard

1:300

Second floor plan (summer)

index
1 dorm
2 storage
3 tridimensional vierscence
4 common classroom
5 terrace
6 platform
7 toilet
8 reading room
9 student activity room
10 hostel for single staffs

1:300

First floor plan (summer)

West elevation

North elevation

应变建筑 climate-responsive building

Sunlight Analysis

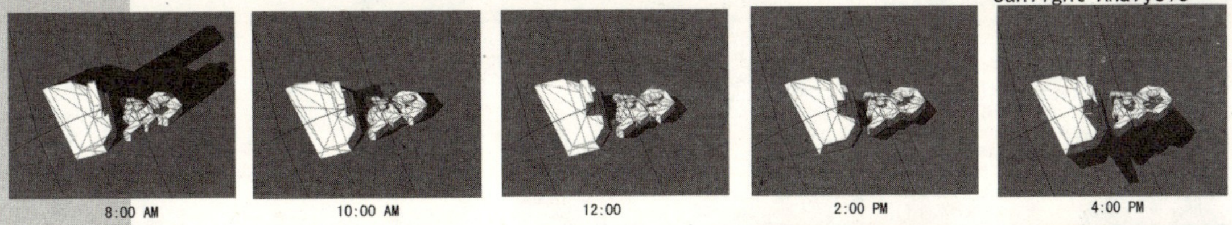

8:00 AM | 10:00 AM | 12:00 | 2:00 PM | 4:00 PM

冬季通过轨道系统将底层架空的围护结构封闭,以蜂穴形式围成一团,场区基本上没有再生风或者二次风现象,而且风速都很低。同时可以有效的减少表面的热损失。从1.5m处场区的风速云图可以看出,场区的最高风速为1.87,远低于标准GB/T50378-2006规定的建筑物周围人行区风速5m/s的要求。同时将食堂四面围合,成为周围建筑的辅助热源。

Through the track system, aerial layer of the ground floor is closed. In Winter, the school building appears like "honeycomb". There is virtually no renewable wind or secondary air, the wind speed is low. At the same time, the buildings close together, so they can effectively reduce thermal loss. The highest wind speed in the height of 1.5m is lower than 1.87m/s, well below 5m/s(the standards provided by GB/T50378-2006). The canteen is combined to become the auxiliary heat source of the surrounding buildings

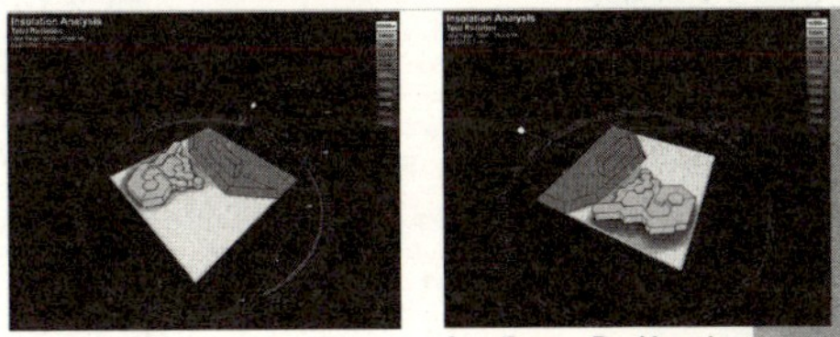

Surface Radiation

Flexible wall

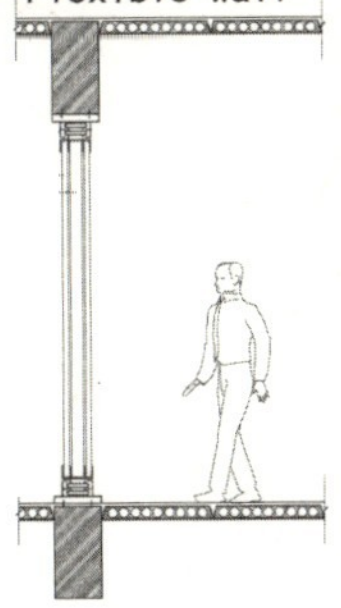

Rain collector

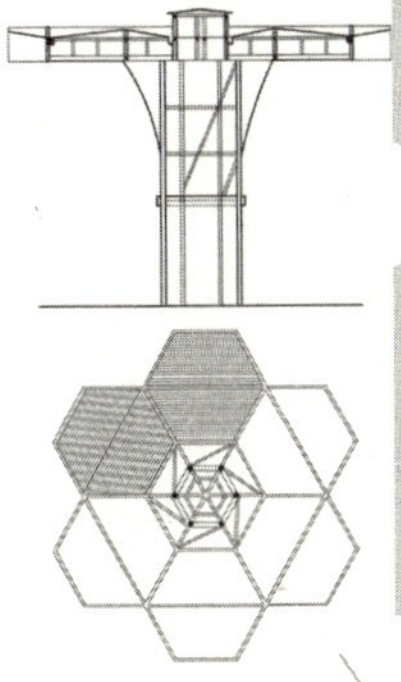

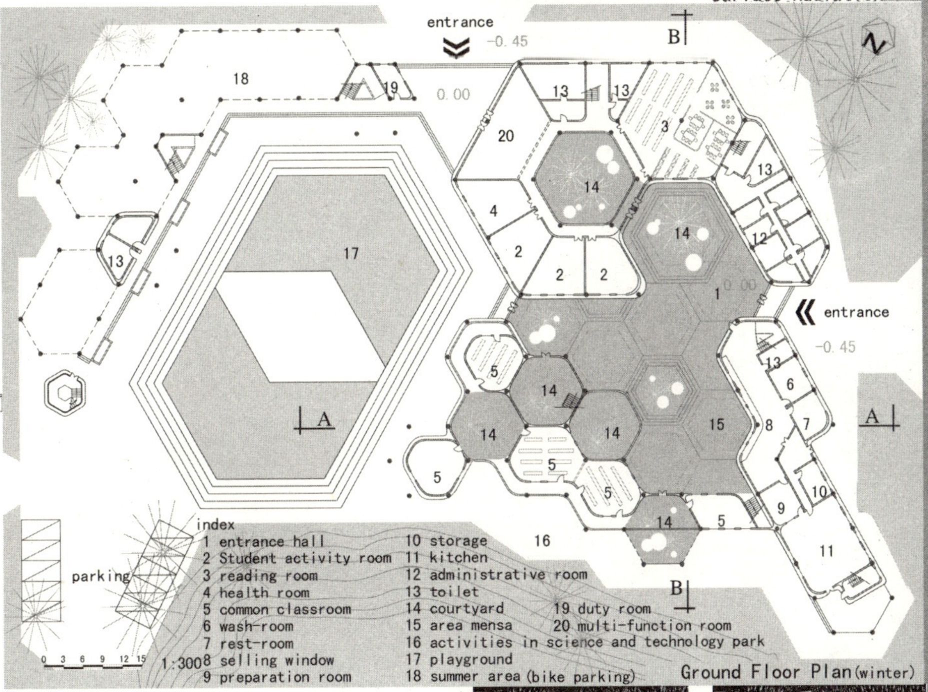

Ground Floor Plan (winter)

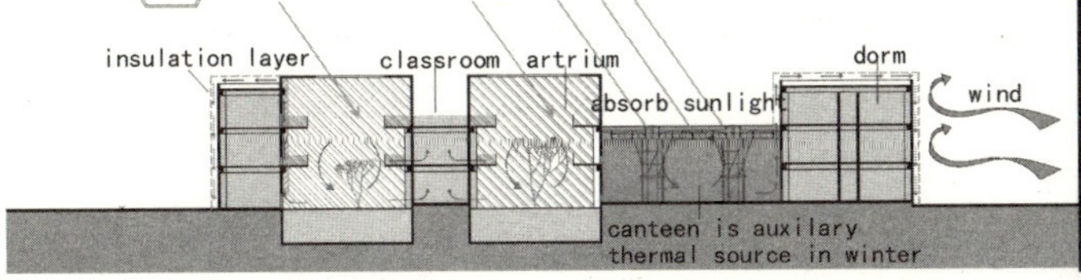

index
1 entrance hall
2 Student activity room
3 reading room
4 health room
5 common classroom
6 wash-room
7 rest-room
8 selling window
9 preparation room
10 storage
11 kitchen
12 administrative room
13 toilet
14 courtyard
15 area mensa
16 activities in science and technology park
17 playground
18 summer area (bike parking)
19 duty room
20 multi-function room

1:300

Section A-A (winter)

insulation layer | classroom | artrium | absorb sunlight | dorm | wind
canteen is auxiliary thermal source in winter

Section B-B (winter)

insulation layer | classroom | absorb sunlight | classroom | rain collector

Ventilation Simulation (winter)

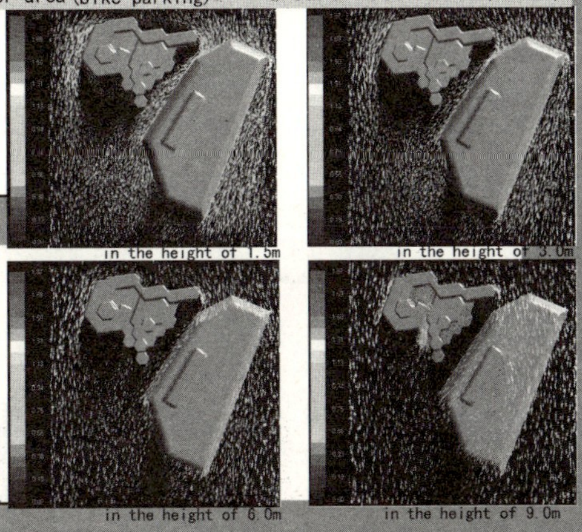

in the height of 1.5m | in the height of 3.0m
in the height of 6.0m | in the height of 9.0m

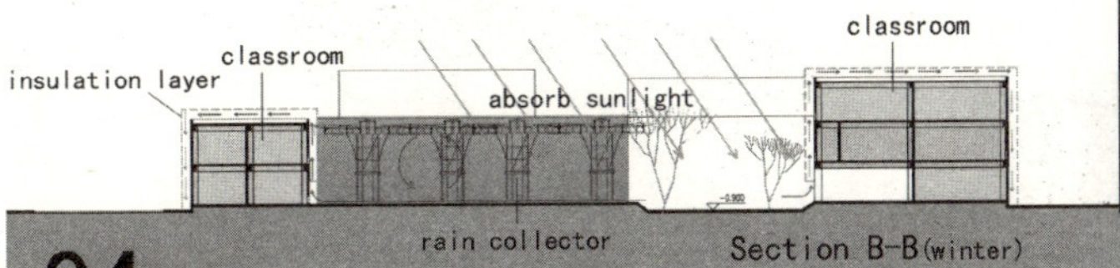

04

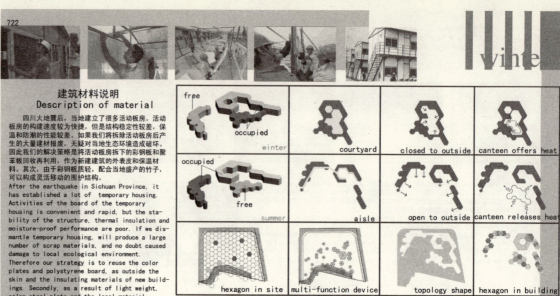

建筑材料说明
Description of material

四川大地震后，当地建立了很多活动板房。活动板房的构建速度较为快捷，但是结构稳定性较差，保温和防潮的性能较差。如果我们将拆除活动板房后产生的大量建材报废，无疑对当地生态环境造成破坏。因此我们的解决策略是将活动板房拆下的彩钢板和聚苯板回收再利用，作为新建筑的外表皮和保温材料。其次，由于彩钢板质轻，配合当地盛产的竹子，可以构成灵活移动的围护结构。

After the earthquake in Sichuan Province, it has established a lot of temporary housing. Activities of the board of the temporary housing is convenient and rapid, but the stability of the structure, thermal insulation and moisture-proof performance are poor. If we dismantle temporary housing, will produce a large number of scrap materials, and no doubt caused damage to local ecological environment. Therefore our strategy is to reuse the color plates and polystyrene board, as outside the skin and the insulating materials of new buildings. Secondly, as a result of light weight, color steel plate and the local material

应变建筑 climate-responsive building

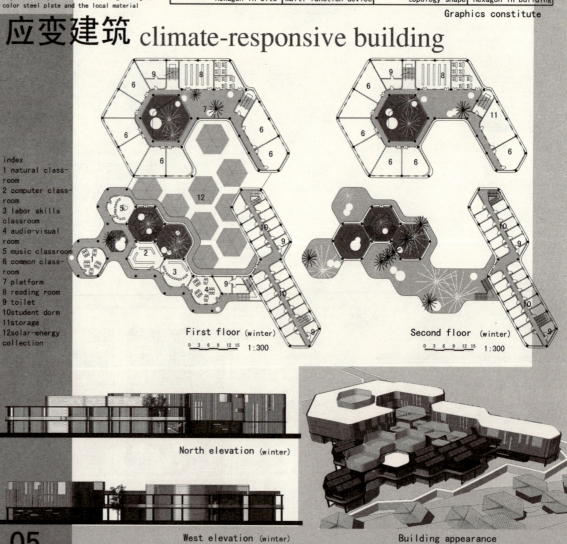

index
1 natural classroom
2 computer classroom
3 labor skills classroom
4 audio-visual room
5 music classroom
6 common classroom
7 platform
8 reading room
9 toilet
10 student dorm
11 storage
12 solar-energy collection

应变建筑 climate-responsive building

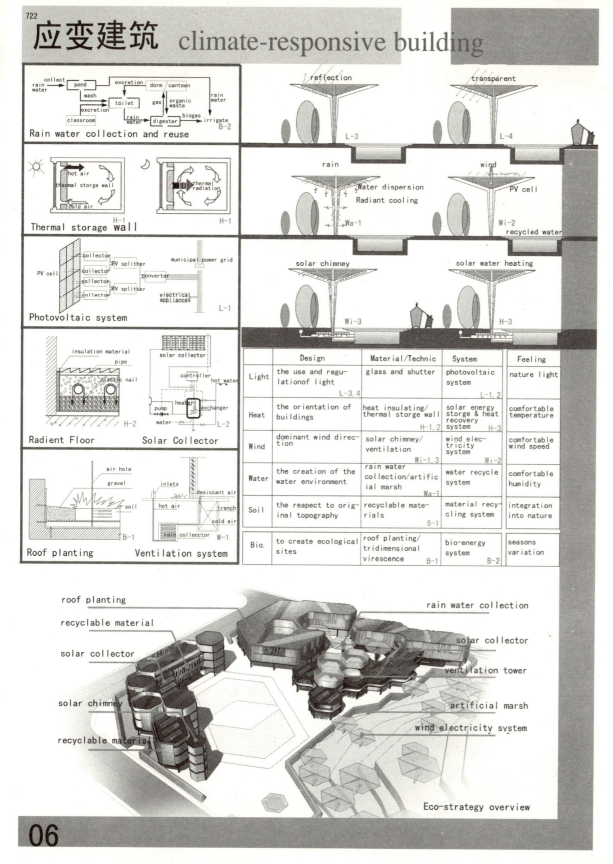

优秀奖
Honorable Mention Prize

项目名称：进化的小学
　　　　　Evolution
作　者：陈维果、王力
参赛单位：重庆大学建筑城规学院

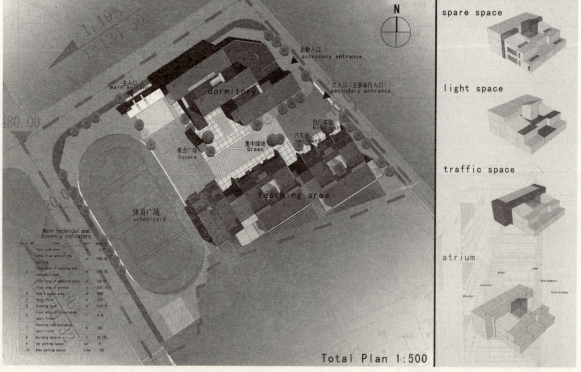

绵阳乡村太阳能小学设计
Sunshine Solar School Design For Mianyang
进化的小学-2 Evolution

ID:733

Technology

Sunshine Greenhouse

1. 作为一个整体调控建筑的一个中转站，通过可调百叶以及植物的选择性载种，调节夏季的阳光直射以及冬季的温度流动。

   The sunshine greenhouse can become a transfer station which is used to control the whole building. It can regulate the sunlight in summer and the temperature in winter through the shutters and the plants.

Corridor

2. 教室连廊的设计充分考虑通风和采光，利用天井和中庭对阴暗的连廊补光，并通过可转动的百叶窗控制风向。

   Classroom full account of the design of the corridor ventilation and lighting, the use of the courtyard and atrium of the dark corridors fill light, and rotating through the shutters to control the wind.

Atrium

3. 顶上盖有透明的玻璃天窗，改善了天井阴暗的状况，使教学楼能达到天然采光。而中庭的拔风效应可以加速建筑内部空气的流动，达到夏季降温的目的。冬天通过玻璃天窗开启角度的调整有效地节省了能耗。

   Top covered with transparent glass skylights, has improved the situation of an open courtyard in the dark, so that building can achieve natural lighting. The atrium effect of pulling the wind can accelerate the flow of air inside the buildings to the summer cooling purposes. Winter through the glass skylight opening angle adjustment effectively save energy.

Vent

5. 在北面教室设置通风口，可接受夏季主导风，形成穿堂风，降低夏季室温；风口较小，避免冬季的热量流失。

   Vents in the north of the classroom setting, acceptable summer winds led to form a draft, lower summer room temperature; tuyere smaller to avoid heat loss in winter.

Roof Planting

4. 教学楼建筑由地面缓缓生成，较大面积的覆土，尽可能和场地融合在一起，弱化建筑的形体感而加强共生的态势，并提供休息的平台和交流空间。

   Teaching building grow up slowly from the ground, the larger area of soil, as far as possible, and venues together, weakening the building's physical sense of the situation and strengthen the symbiotic and provide a platform for rest and spare space

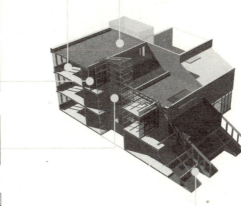

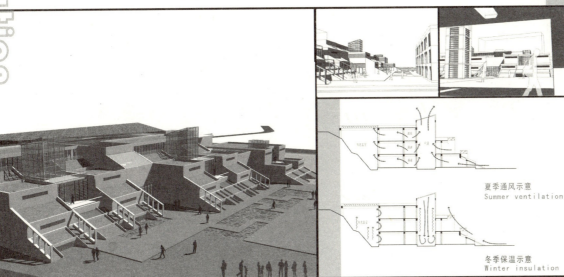

夏季通风示意
Summer ventilation

冬季保温示意
Winter insulation

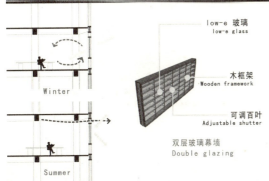

low-e 玻璃
low-e glass

木框架
Wooden framework

可调百叶
Adjustable shutter

双层玻璃幕墙
Double glazing

Winter

Summer

建筑的南立面采用双层表皮玻璃幕墙，其中外层幕墙采用双层真空玻璃，可以打开通风，内层则是9毫米厚的单层玻璃。框架采用木材，杜绝冷桥。两层幕墙中间装有百叶窗，用于调节进光亮，冬季室外空气在进入室内前，先在两层玻璃间预热，保证室内的舒适度；而在夏季，可以将外层幕墙打开并放下百叶窗，减少进入室内的太阳辐射，同时进行通风。

Construction of the Southern double-skin facade of glass walls, of which the outer walls of double-vacuum glass, can turn on the ventilation, the inner layer is 9 millimeters thick single-layer glass. Framework for the use of wood, put an end to the cold bridge. Intermediate walls with a two-tier shutters for the regulation into the bright winter outdoor air into the room, before a two-tier glass in between warm-up to ensure that the indoor comfort; while in the summer, you can open the outer walls and lay down their shutters. Reducing access to the indoor solar radiation, at the same time ventilation.

绵阳乡村太阳能小学设计
Sunshine Solar School Design For Manyang
进化的小学-3 Evolution

ID:733

Teaching area   1-1 section 1:300

Skyline

Tower as a whole composition Center site so that schools from all directions to see it had very beautiful skyline, and the vertical and horizontal to the tower of the teaching building and hostels have a strong contrast to the construction site a more unified group.

Void and Entity

Glass courtyard in addition to its own function, the overall composition in the construction also has a key role. The use of vitreous interspersed contrast actual situation caused by the construction and dialogue with the environment. While the actual situation changes in the construction unit also strengthened the sense of rhythm.

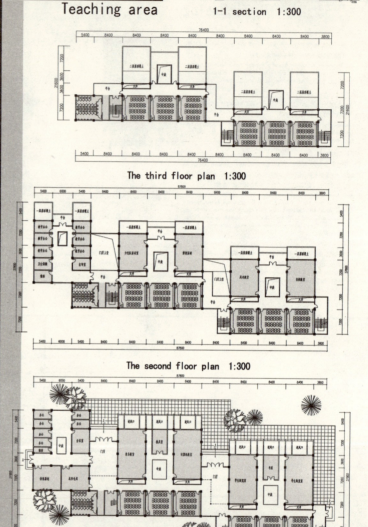

The third floor plan 1:300

The second floor plan 1:300

The first floor plan 1:300

West elevation 1:300   North elevation 1:300

East elevation 1:300

绵阳乡村太阳能小学设计
Sunshine Solar School Design For Mianyang
进化的小学-4 Evolution

ID:733

Dormitory area

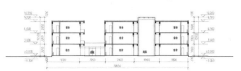

1-1 section 1:300

Ventilation Path

Venues summer led to the Northwest wind, so towards the south east as to maximize the use of the summer wind driven. Teaching building and the layout of houses have also done a way conducive to guide the direction of the processing – will dominate the wind into the wind as far as possible designed to open Office, and the relative austerity exports and accelerate the wind flow at the venue.

shutter

Dormitory of the adjustable settings Venetian walls, to stop the winter wind and To promote the ventilation in summer. In addition, planted in the thermometer screen above a certain Tetrastigma plants, building facades and roofs have echoed the relationship between soil.

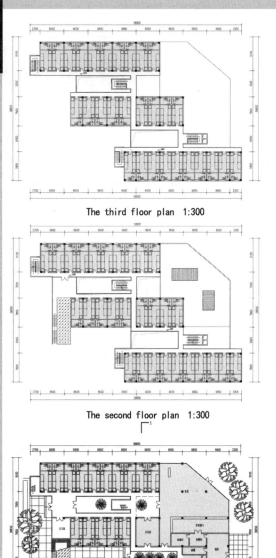

The third floor plan 1:300

The second floor plan 1:300

The first floor plan 1:300

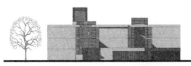

East elevation 1:300

South elevation 1:300

West elevation 1:300

North elevation 1:300

优秀奖
Honorable Mention Prize

项目名称：光阴趣事
　　　　　Sunshine Shadow & Fun
作　　者：李荣、吕家悦、谢崇实、陈潇、
　　　　　吕晓田、杨鹏程、黄一滔、
　　　　　何兆熊、冷冷、陈佐球
参赛单位：重庆大学建筑城规学院、城市建设
　　　　　与环境工程学院

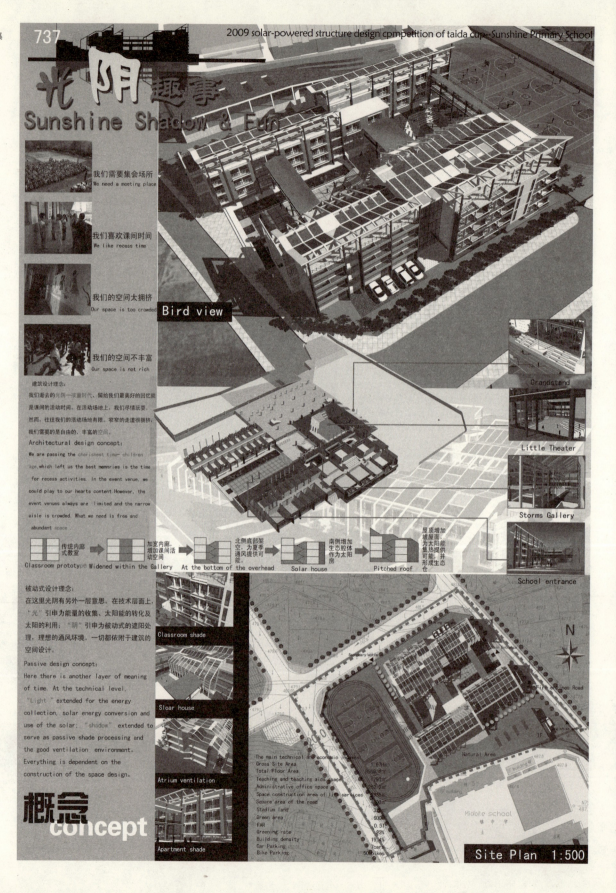

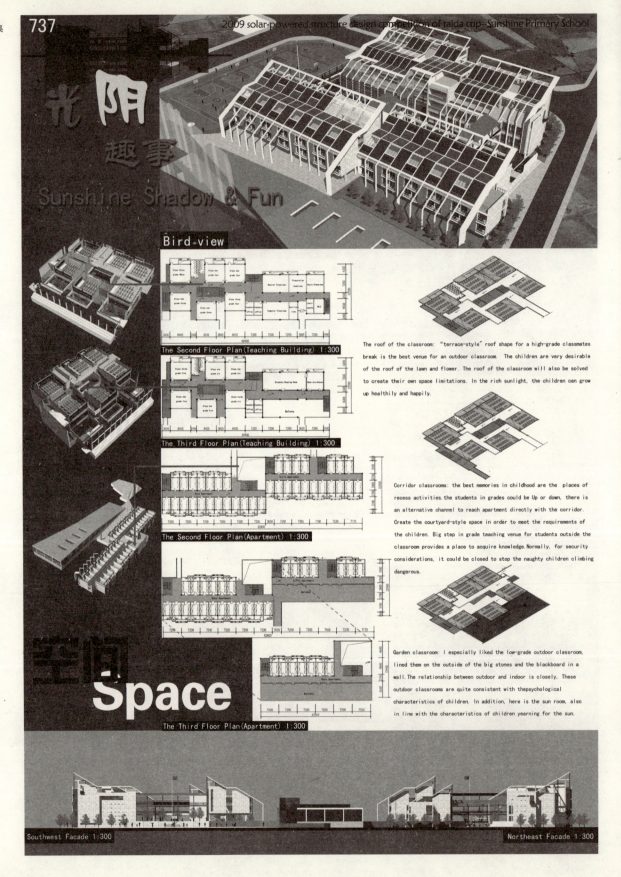

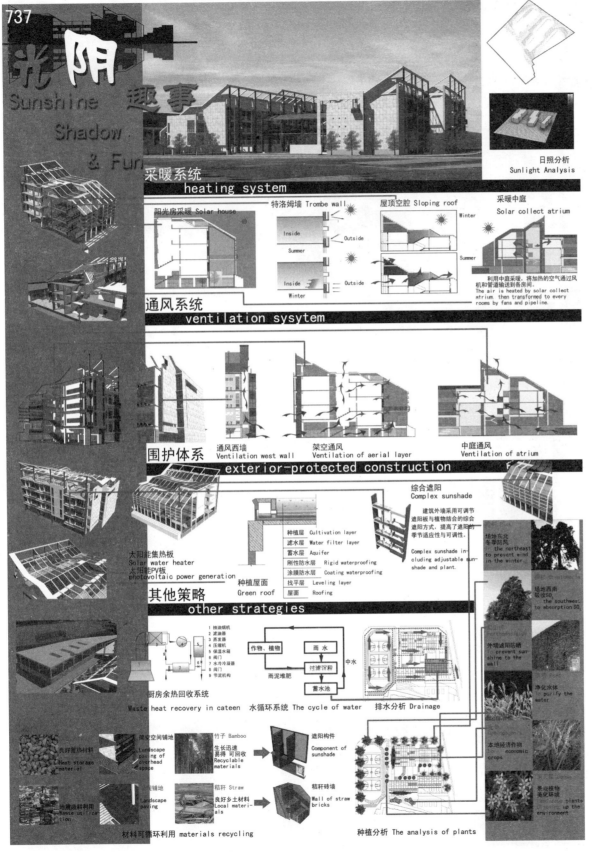

优秀奖
Honorable Mention Prize

项目名称：光·盒
Sunshine & Square

作　　者：高力强、邓可祥、刘瑞杰、欧阳文、
　　　　　黄帅、刘丹、丁磊、朱江涛

参赛单位：石家庄铁道学院建筑与艺术分院、
　　　　　河北工业大学

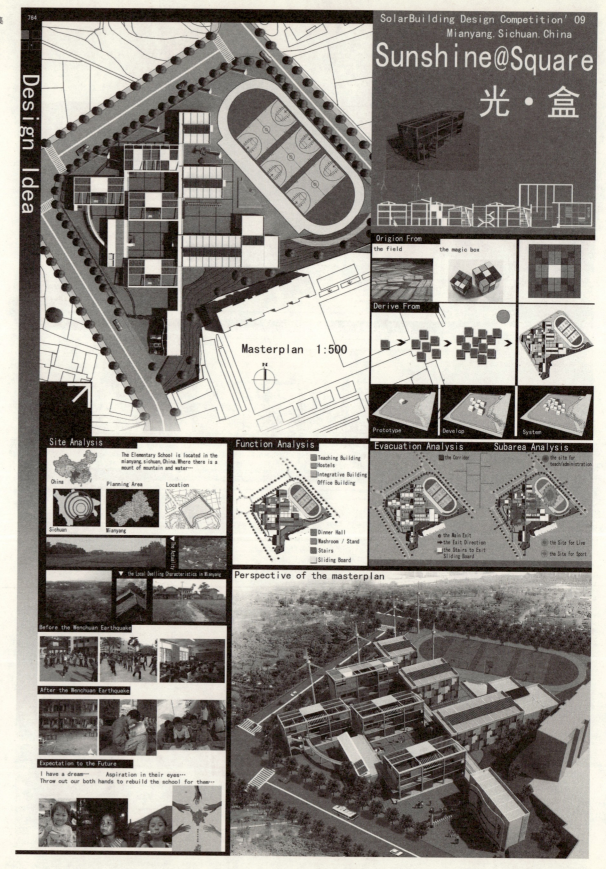

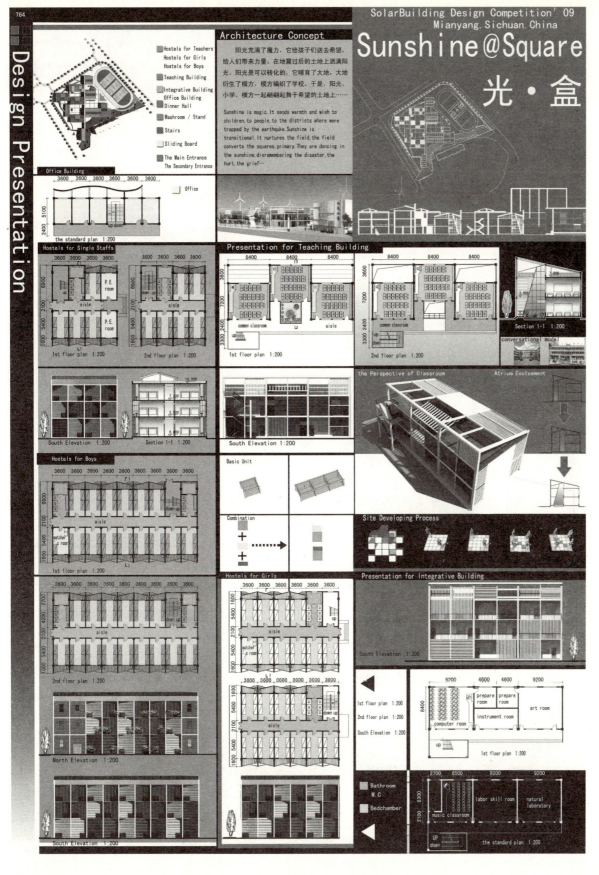

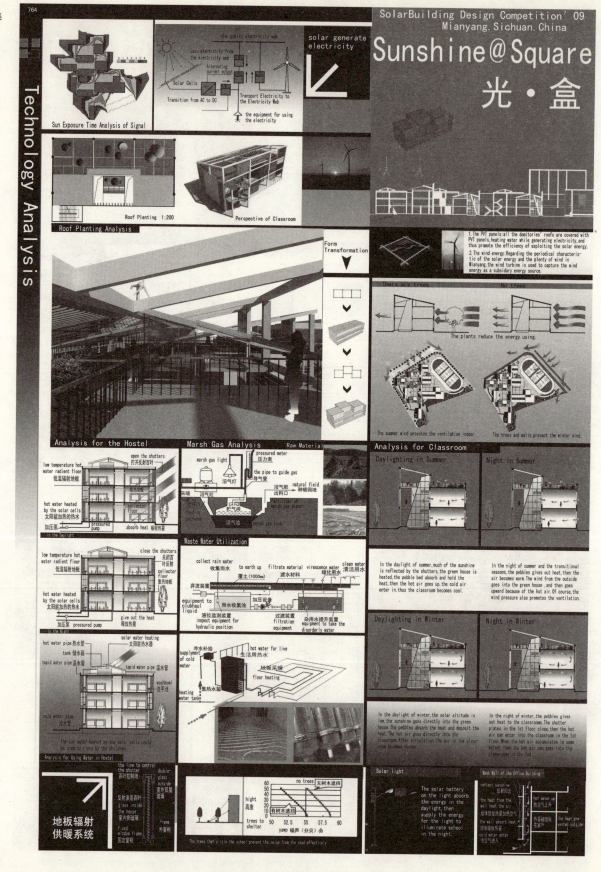

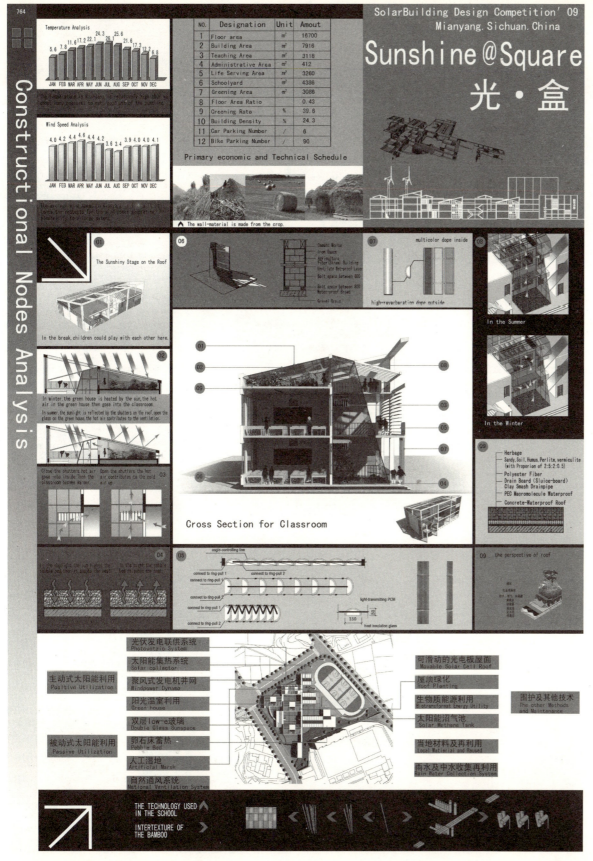

优秀奖
Honorable Mention Prize

项目名称：时光 阳光 童年
Time Sunshine Childhood
作　者：张之光、康冬、薄超、宫月
参赛单位：山东建筑大学建筑城规学院

130

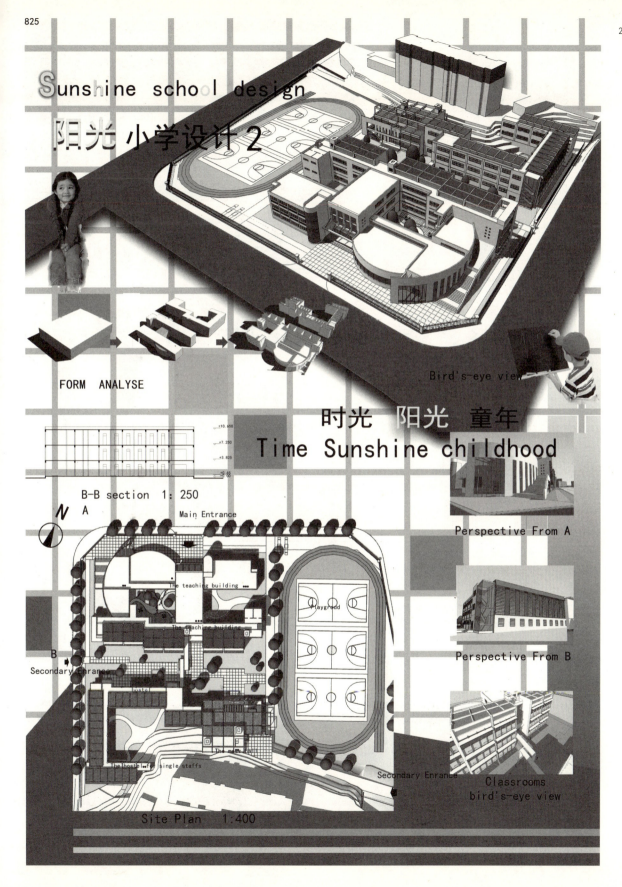

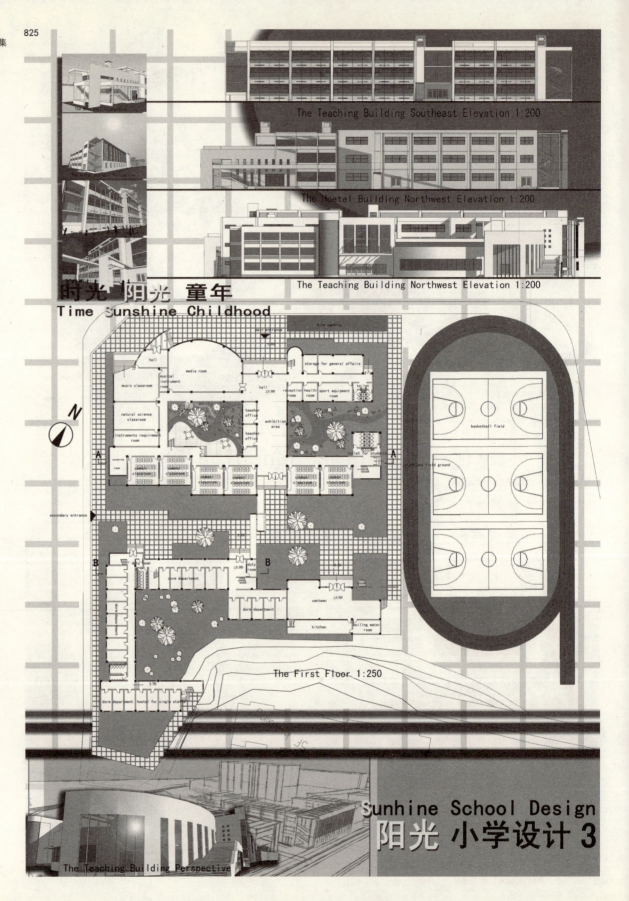

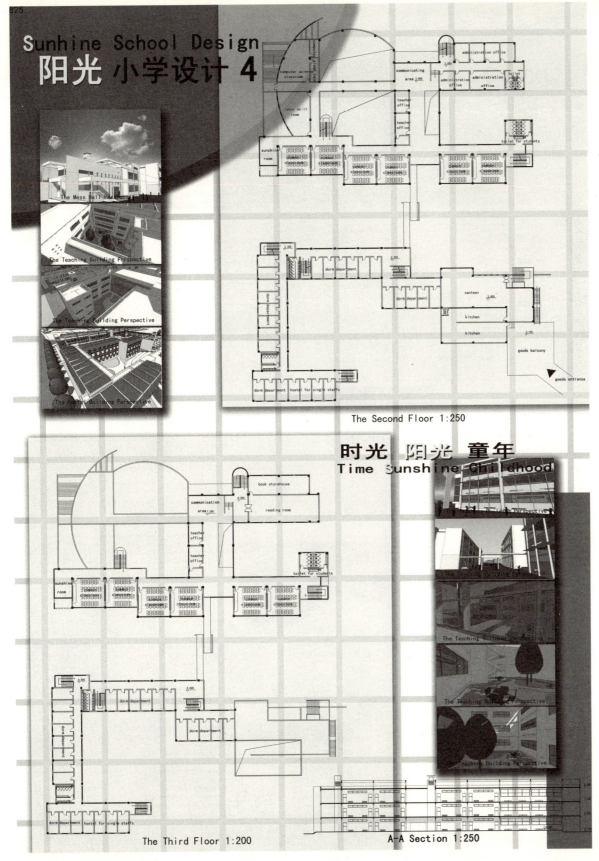

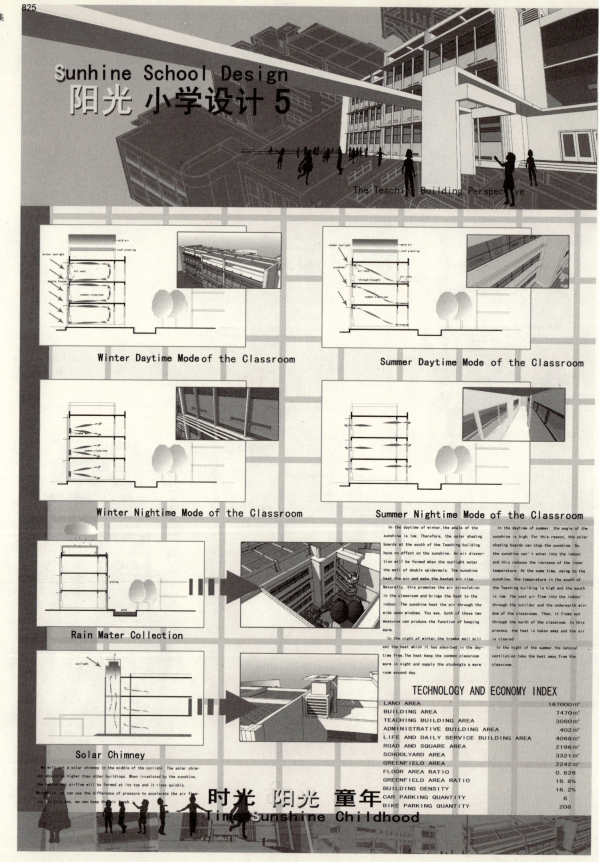

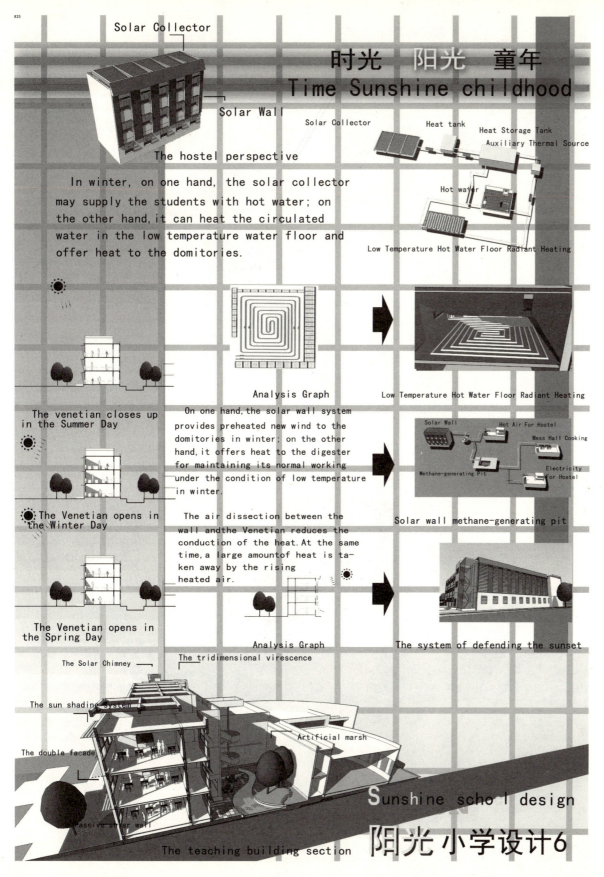

2009台达杯国际太阳能建筑设计竞赛获奖作品集

优秀奖
Honorable Mention Prize

项目名称："L"代表什么？
　　　　　What is "L"？
作　　者：焦尔桐、徐绍辉、张文超、
　　　　　黄砂、刘中爽、韩雪
参赛单位：山东建筑大学建筑城规学院

For the May 12th, 2008 earthquake in southern Sichuan, the nineteenth deadliest earthquake of all time, the 2008 Sichuan earthquake, which measured at 8.0 Ms occurred at 14:28:01.42 CST on May 12, 2008 in Sichuan province of China and by any name killed at least 69,000 less than three months before China hosted the world in the 2008 Summer Olympics.

The central government estimates that over 7,000 inadequately engineered schoolrooms collapsed in the earthquake.

School — WHAT IS "L"? — LOCALITY

设计是人文理想	Design is humanity ideal
设计是社会责任	Design is social responsibility
设计是生活的原动力	Design is the power of life
设计协调矛盾并解决问题	Design can coordinate the contradiction and solve a problem
阳光与希望，让设计返璞归真	Sunlight and hope, let design return its original simplicity
————设计呵护生命	
优良的设计不需要庞大的规模和奢华的配置也不必以维护它的运营成为师生日后的负担	Good design does not need ample scale and luxurious allocation
学校的规划简约而自然	The school plan is concise and natural
学校的建筑安全而朴素	The school building is simple but safe
学校的教学环境温暖舒适又充满关怀	The school teaching environment is full of warm comfort and shows solicitude
学校的景观既有地域性特征又结合自然设计	The school landscape have the regionality characteristic with natural design combined
能源利用系统的高效率符合环保的要求又考虑长远发展	The source of energy makes use of systematic high efficiency to accord with environmental protection's demanding to consider long-term development
设计的点滴只为孩子们能够和着阳光与希望的旋律茁壮成长	The intravenous drip designing is able to be preparing sunlight and the melody hoping for the children's growing up

Design the subject and call for Design that being productivity, is life-style, being community responsibility design is today solve a problem, Servicing designing that design that this is one ah at all protecting life most average return county level elementary school does not require that super-big scale and luxurious allocation do not dodge to defend it once again becomes teacher and student in the future bearing building designing that common contributes to her down once of boundary business circles invests the sufficient plan to place school in construction is intensive but natural school in Guangdong is that simple but safe school interior environment, simple kindness of equipment are full of human nature showing solicitude for the school landscape being the common people, ego Leading growing looks at the system image limpid boundless children of innovative idea is penetrating the perfectly fit portable school uniform being humming self happy grow up if the disaster approaches to have emergent passage, safe song route, the haven, the emergency are illuminating, saving self to be left all to, ...

MianYang Sunshine Primary School Design   HOPE & SUNSHINE 01

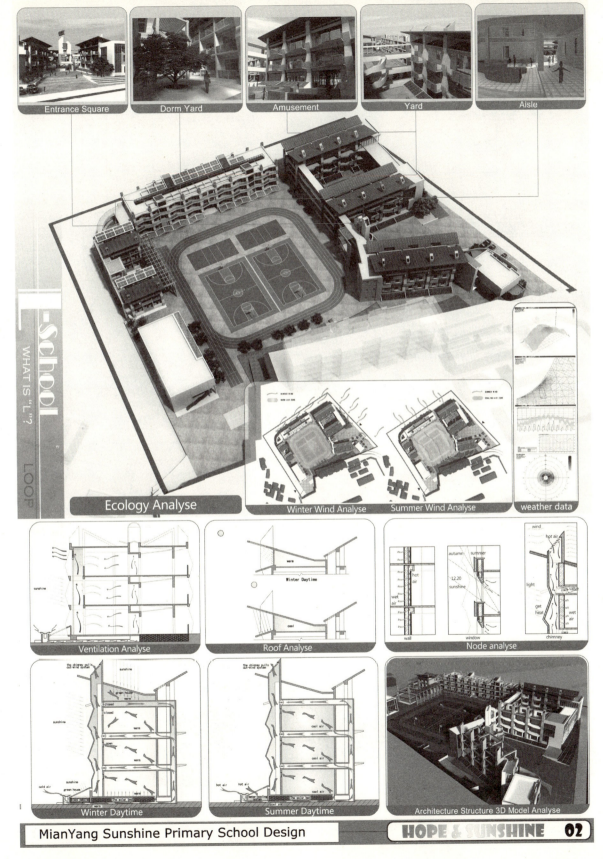

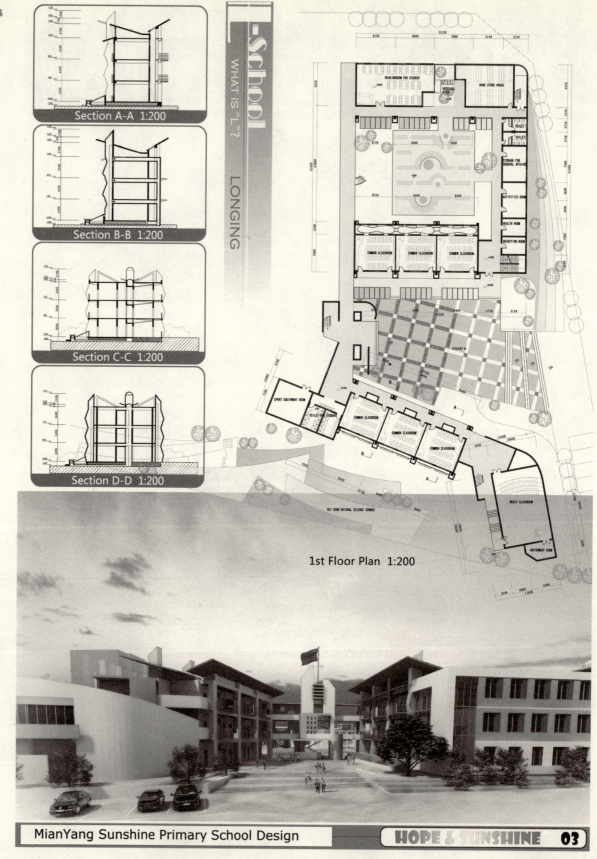

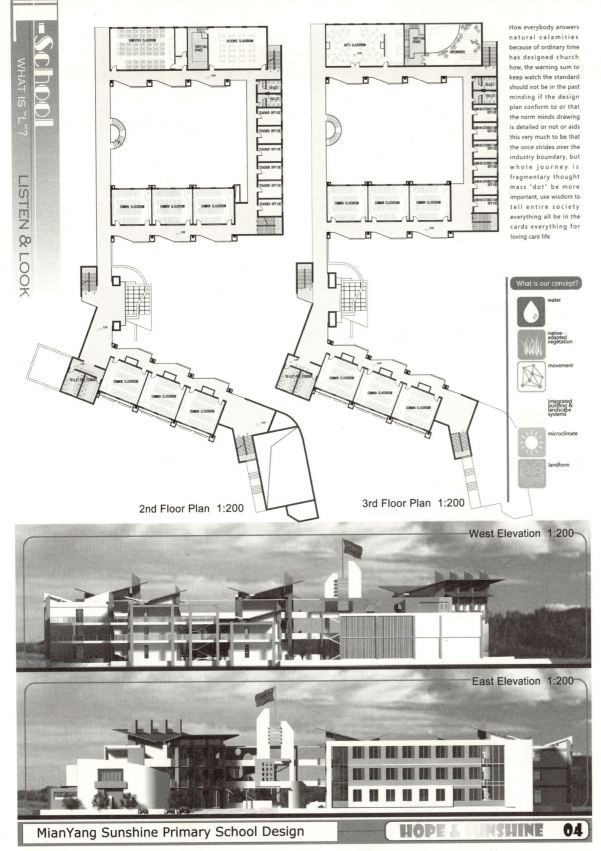

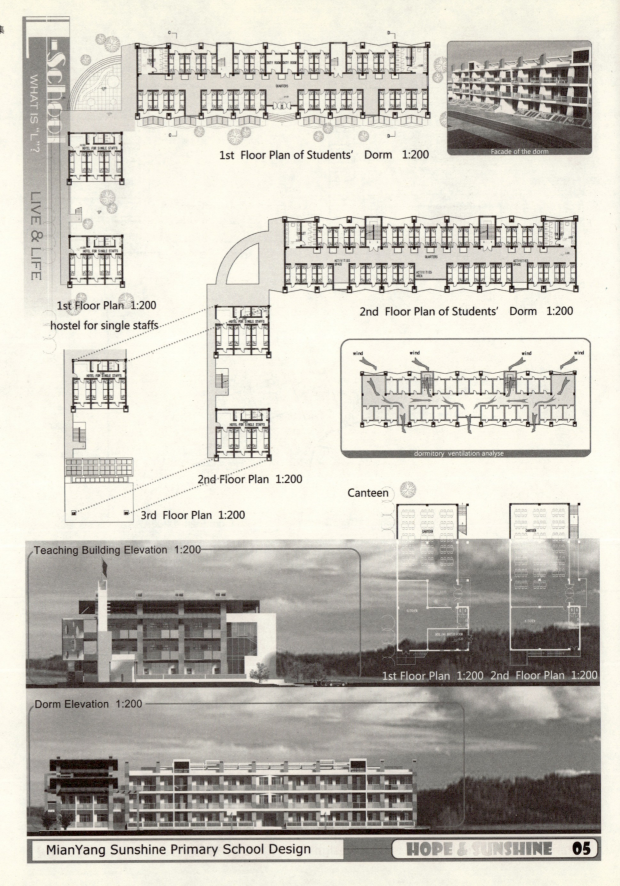

MianYang Sunshine Primary School Design

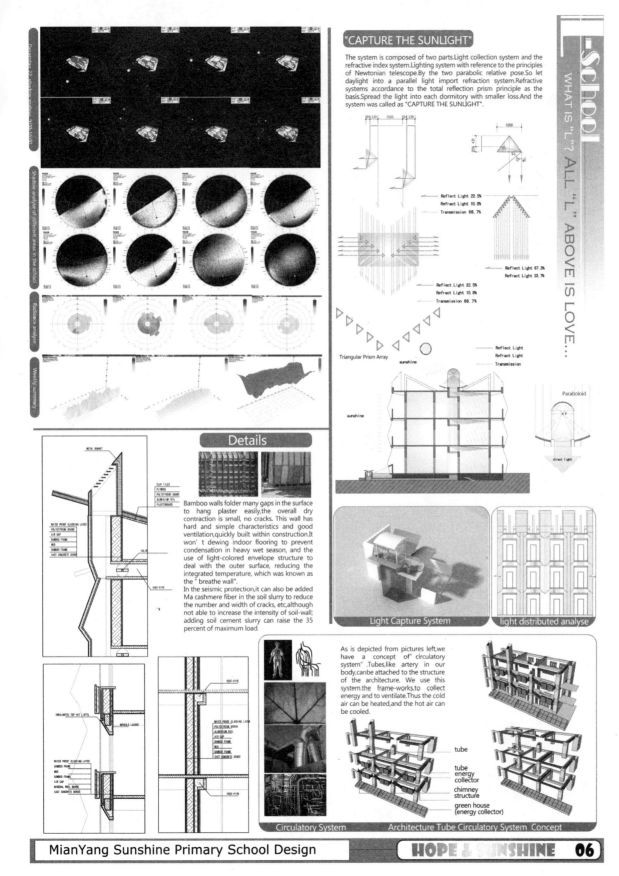

优秀奖
Honorable Mention Prize

项目名称：希望 绿色 生态
　　　　　Hope Green Ecologica
作　　者：司鸿斌、王卫超、王博成、李静
参赛单位：山东建筑大学建筑城规学院

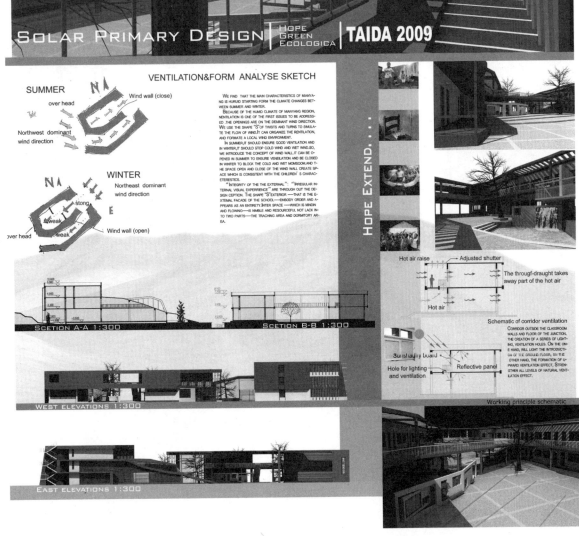

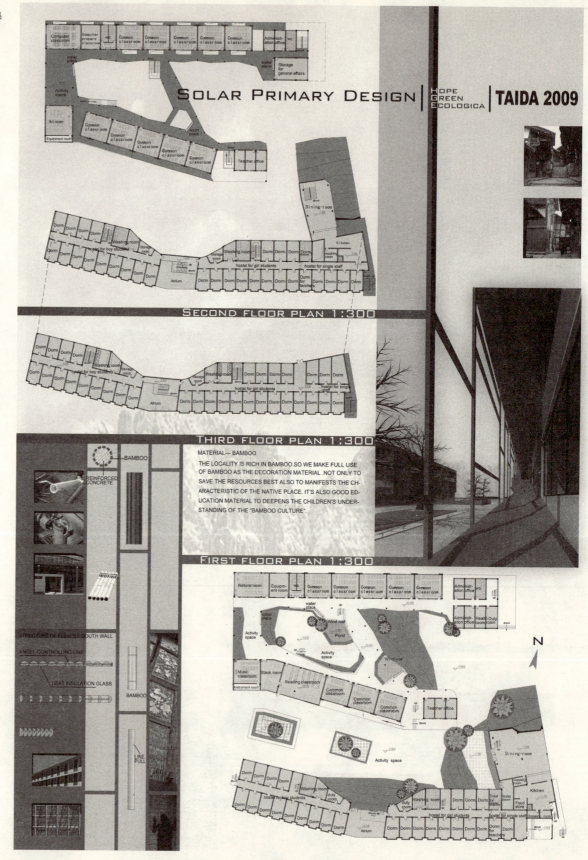

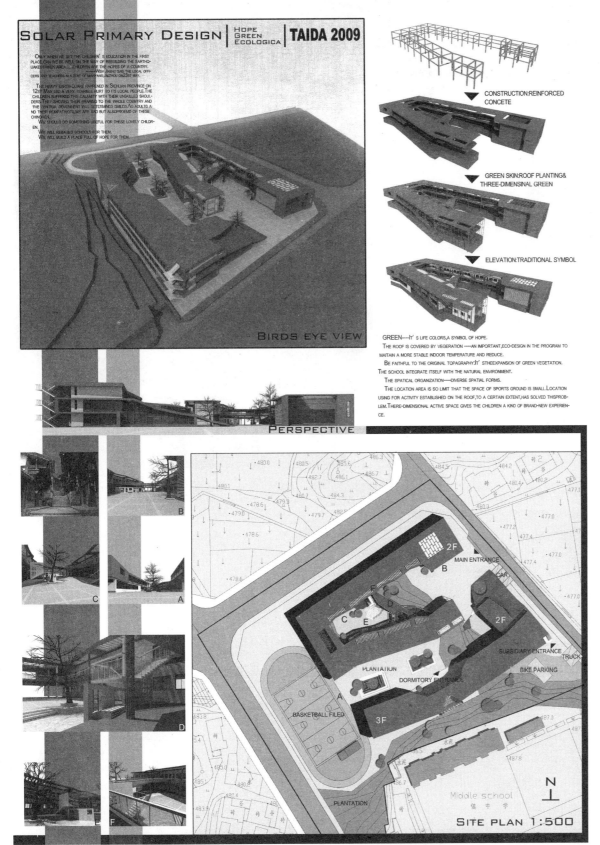

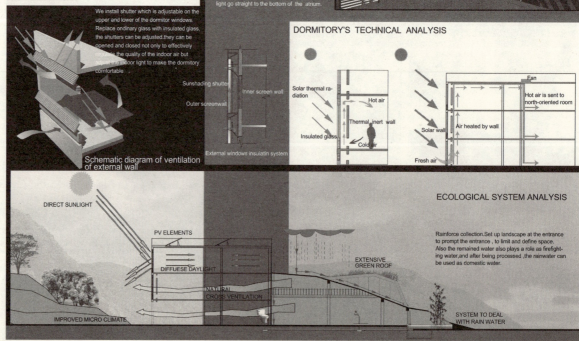

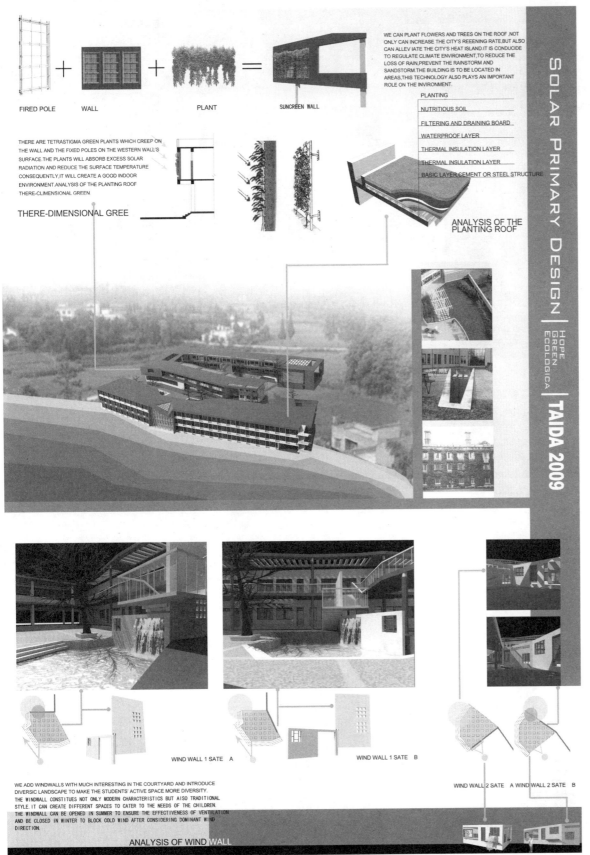

优秀奖
Honorable Mention Prize

项目名称：阳光"家"希望
　　　　　Sun+Hope=Home
作　　者：曲羽、李楠、李崭、朱晓松
参赛单位：山东建筑大学建筑城规学院

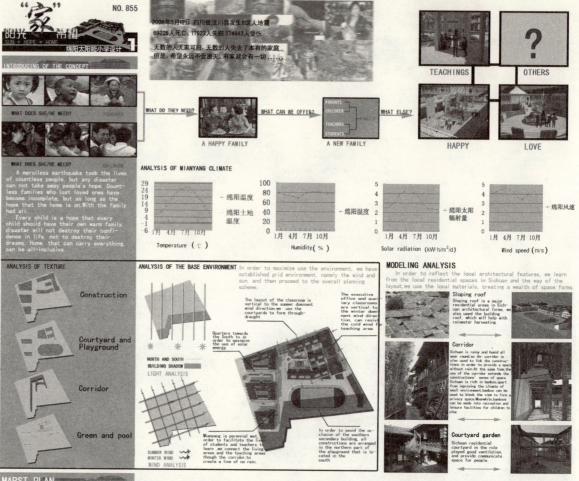

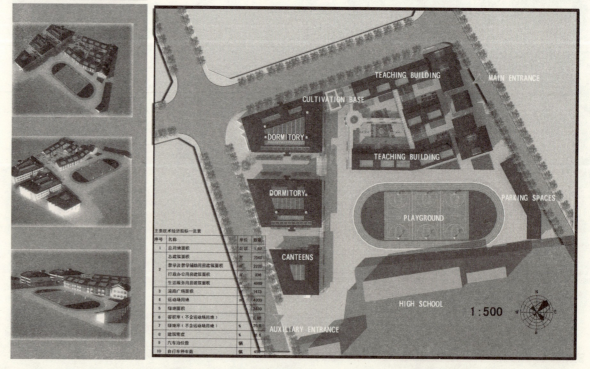

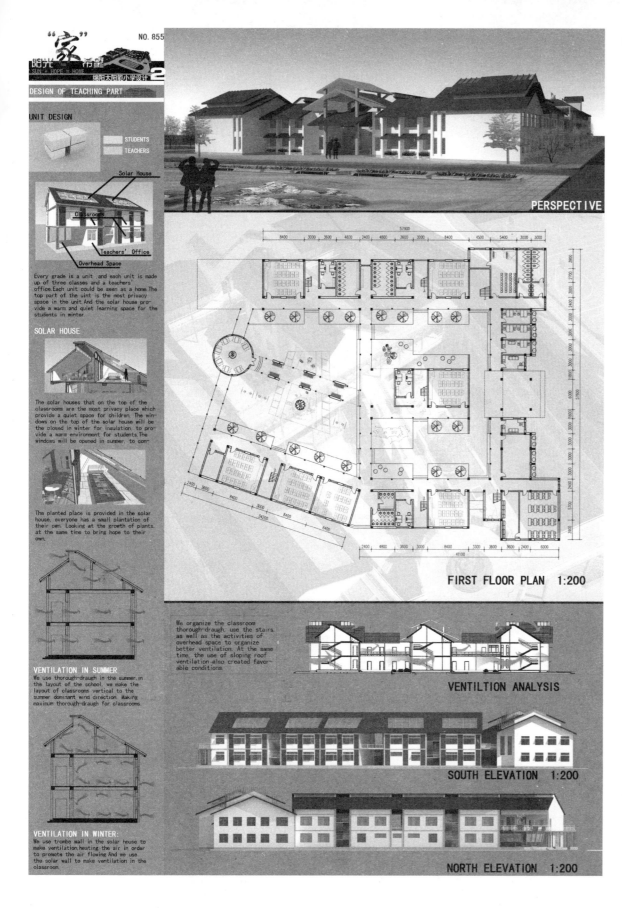

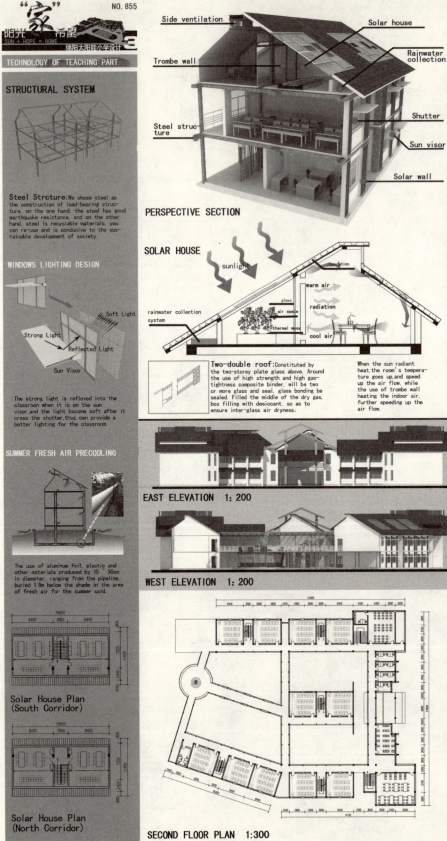

DESIGN OF LIVING PART

UNIT DESIGN

The feeling of home, design of the Dormitory, we make each layer as a unit, teachers' hostels and students' hostels are combination layout, and establish a activity room similar to a family, a room to facilitate the exchange between teachers and students, the family-like layout of the form inorder to the give them the warm feelings of their home. Atrium can be grown plants, regulate the indoor environment and can create a comfortable and natural atmosphere

SOLAR HOUSE

The same as classrooms, we set the solar house for the students on the top of the dormitories, and it provide a quiet and warm learning environment.

ATRIUM SPACE

In order to reduce heat loss, we choose double-sided arrangement. However, taking into account the two-sided arrangement is not conducive to the lighting and ventilation, we set up the atrium space, not only solved the problem of lighting and ventilation, also created a greenhouse space.

SEPARATE ENTRANCE

We set up a separate entrance and stairs for teachers in the dormitories. It is convenient for the teachers to communicate and it also enriched the west elevation of the dormitories.

RESIDENTIAL UNIT DESIGN

In the design of teachers' dormitories, we have adopted residential unit design ways. Provide the teachers a living space with two besrooms and a living room. This design way give privacy to the teachers and also creat better conditions for the communications between the teachers themselves and with the students.

PERSPECTIVE

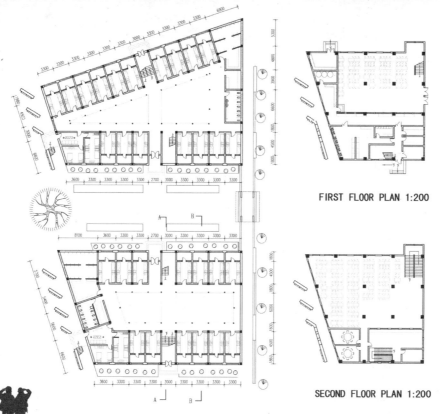

FIRST FLOOR PLAN 1:200

SECOND FLOOR PLAN 1:200

FIRST FLOOR PLAN 1:200

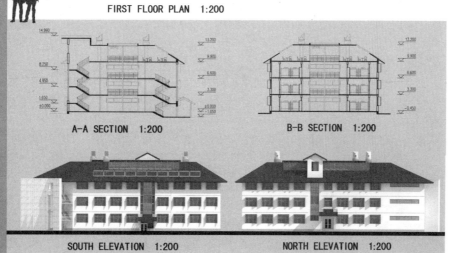

A-A SECTION 1:200

B-B SECTION 1:200

SOUTH ELEVATION 1:200

NORTH ELEVATION 1:200

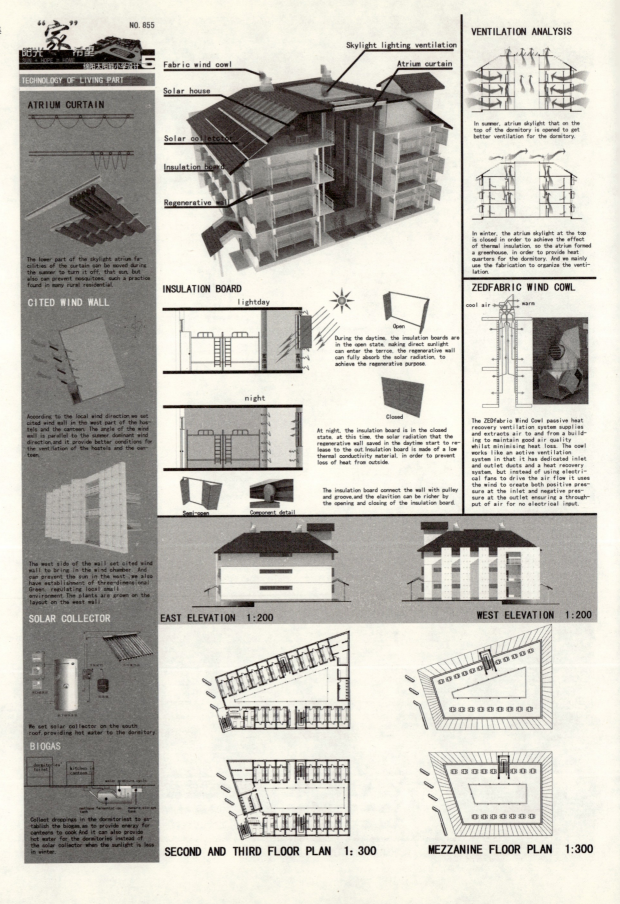

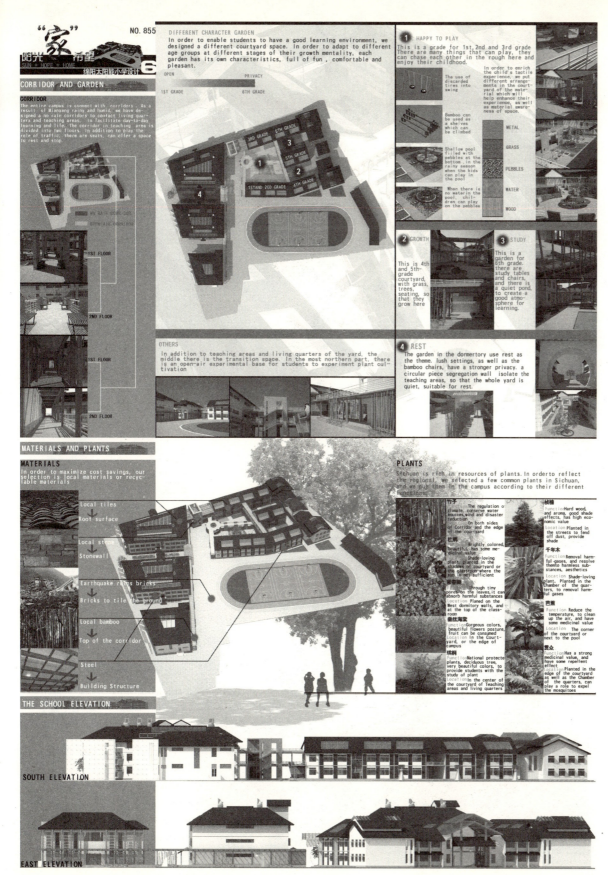

优秀奖
Honorable Mention Prize

项目名称：日光播放器
　　　　　Solar Player
作　　者：石峰、李敏、彭鹏
参赛单位：华中科技大学建筑与城市规划学院

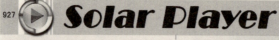

Introductions:

The Mianyang Yangjiazhen Solar Primary School is located in the hot summer and cold winter area. This design respects the zonal characteristics in Mianyang district, and carries out the sustainable design concept. With the combination of the theme of rebuilt after the earthquake, the school brings sunshine and hope to the children. The design adopts the eco-design straytegy with passive solar energy utilization based, combining with the active solar energy utilization, and takes natural ventilation for supplement. The eco-atrium is the key of the entire design, which works as the 'lung' adjusting the microclimate indoor. The Passive Downdraught Cooling Tower is a complement to the eco-atrium, which enhances natural ventilation, and reduce energy consumption.

设计说明：绵阳杨家镇阳光小学地处夏热冬冷地区，设计尊重绵阳地域特点，充分贯彻可持续发展的设计观，，并结合灾后重建的主题，带给灾区儿童以阳光与希望。注重绿色生态技术特别是太阳能技术的应用，采用以被动式太阳能利用为主、结合主动式太阳能利用及自然通风为辅的生态设计策略。生态中庭是整个设计的核心，作为调节室内小气候的"肺"，巧妙化解夏热冬冷地区的种种气候矛盾。捕风井作为生态中庭的强有力补充，可以强化自然通风，降低能耗。

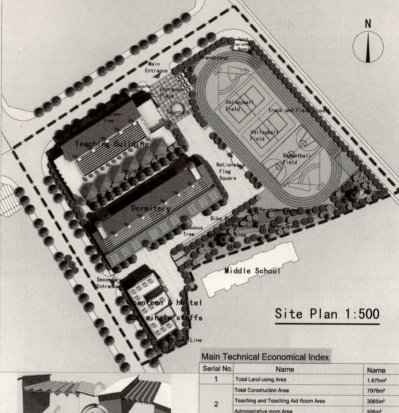

Site Plan 1:500

With the adjusting of the atrium shutter, sunshine of different quantity are introduced. It is just like a media player which you can control when you are enjoying the music.

Main Technical Economical Index

Serial No.	Name	Name
1	Total Land-using Area	1.67hm²
2	Total Construction Area	7976m²
	Teaching and Teaching Aid Room Area	3065m²
	Administrative room Area	306m²
	Domestic Room Area	4605m²
3	Square and Road Area	4802m²
4	Playground Area	3610m²
5	Greening Area	4785m²
6	Floor Area Ratio（not including the playground）	0.48
7	Green Coverage Rate（not including the playground）	28.7%
8	Building Density	21.0%
9	Car Parking Number	6
10	Bike Parking Number	67

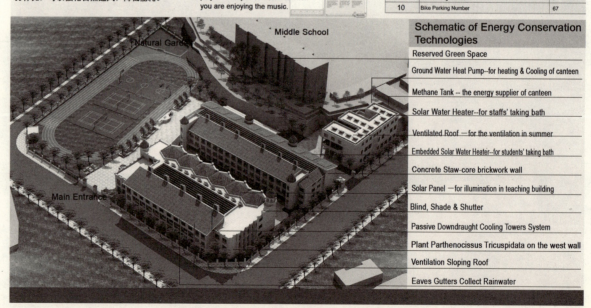

Schematic of Energy Conservation Technologies

- Reserved Green Space
- Ground Water Heat Pump--for heating & Cooling of canteen
- Methane Tank -- the energy supplier of canteen
- Solar Water Heater--for staffs' taking bath
- Ventilated Roof —for the ventilation in summer
- Embedded Solar Water Heater--for students' taking bath
- Concrete Staw-core brickwork wall
- Solar Panel —for illumination in teaching building
- Blind, Shade & Shutter
- Passive Downdraught Cooling Towers System
- Plant Parthenocissus Tricuspidata on the west wall
- Ventilation Sloping Roof
- Eaves Gutters Collect Rainwater

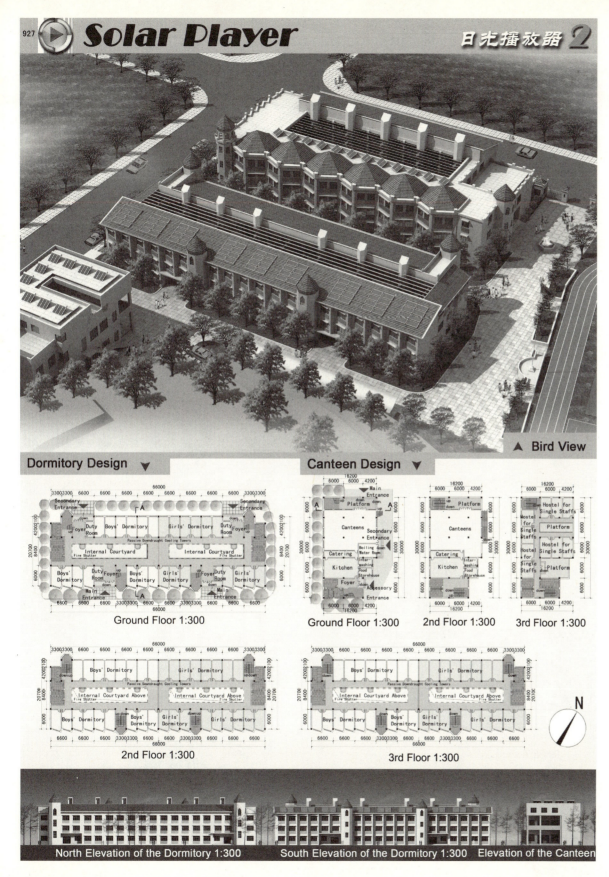

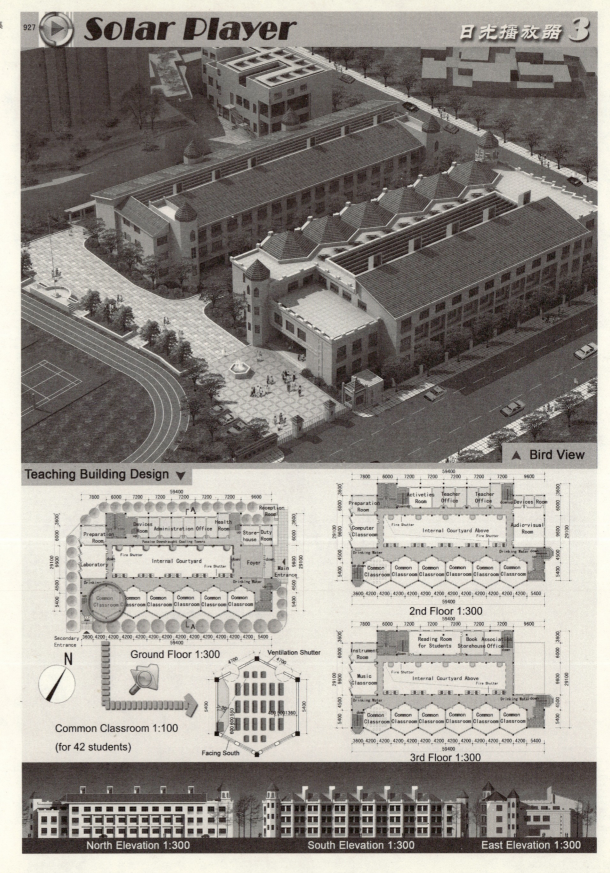

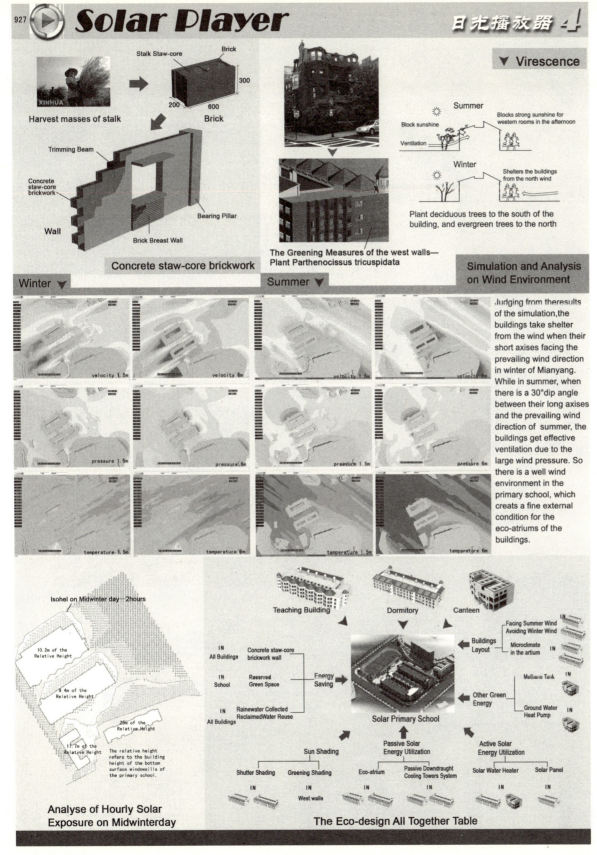

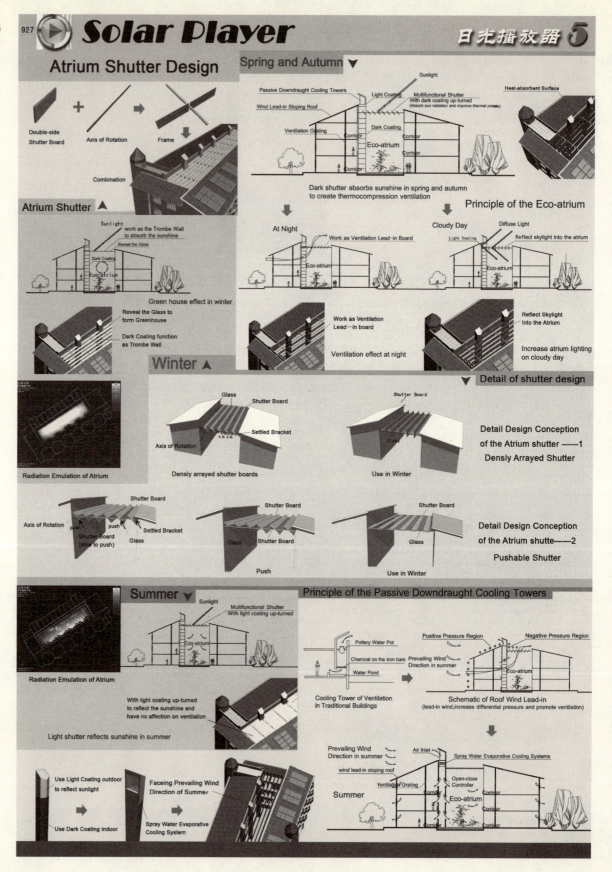

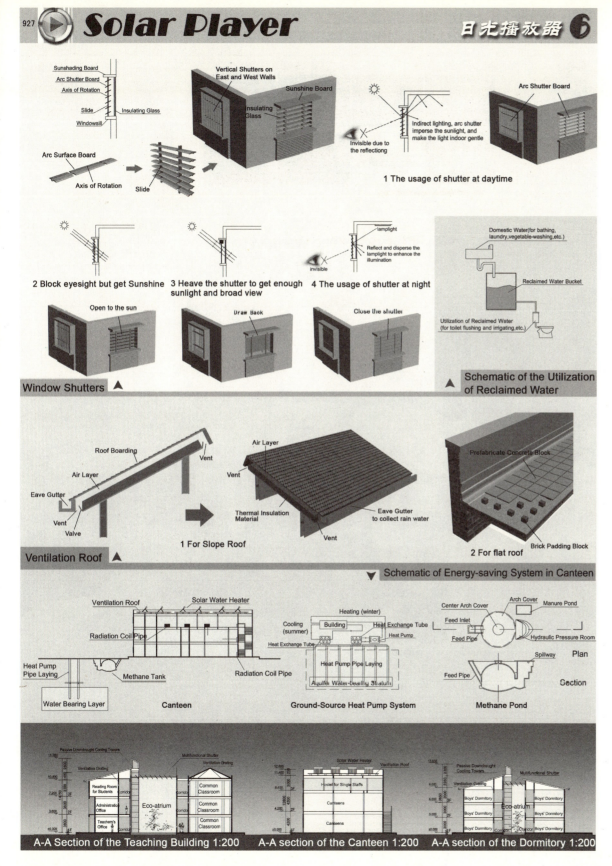

2009台达杯国际太阳能建筑设计竞赛获奖作品集

THE BUFFER BOX | BASIC ISSUES & BASIC CONCEPT
Mianyang Eco-School Design
ID: 1052

优秀奖
Honorable Mention Prize

项目名称：缓冲盒
The Buffer Box
作　者：王飞、郑凯竞
参赛单位：清华大学建筑学院建筑系

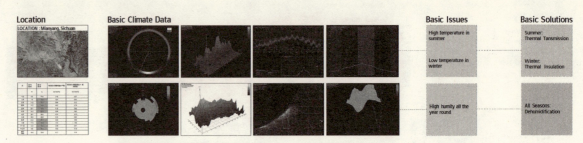

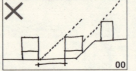

There is a building on the raised ground off the south of the site, if the new school building is located in the south of the site, the shadow area will cover the building.

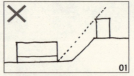

Considering the shadow area's effect, the new school building should be located in the north of the site and most functions are managed in a single box, but spaces facing the north are not ecological.

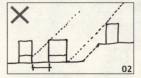

If we seperate spaces into boxes and locate them off the shadow area made by the building on the raised ground, the distance between buildings is too small to get them enough light.

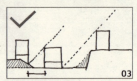

Above all, our solution is to "cut" the earth in the south of the site and move it to the north, to raise the building in the north, which can help it get more light, at the same time, we can get more space in the north for children's outdoor activity.

THE BUFFER BOX | ECO DETAILS & CONCEPT PROCESS
Mianyang Eco-School Design
ID: 1052

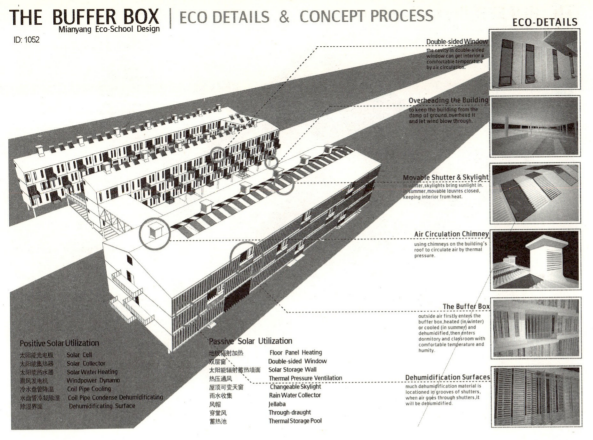

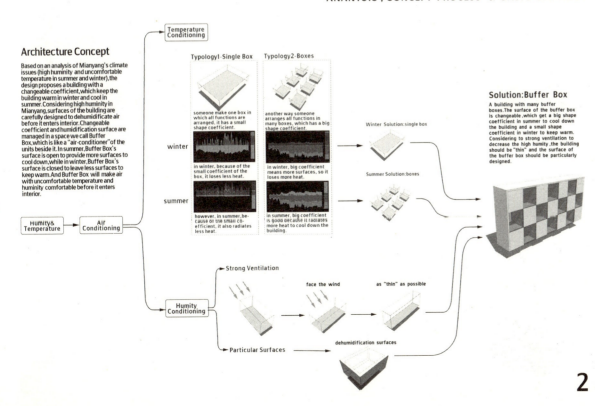

THE BUFFER BOX | SITE PLAN
Mianyang Eco-School Design
ID: 1052

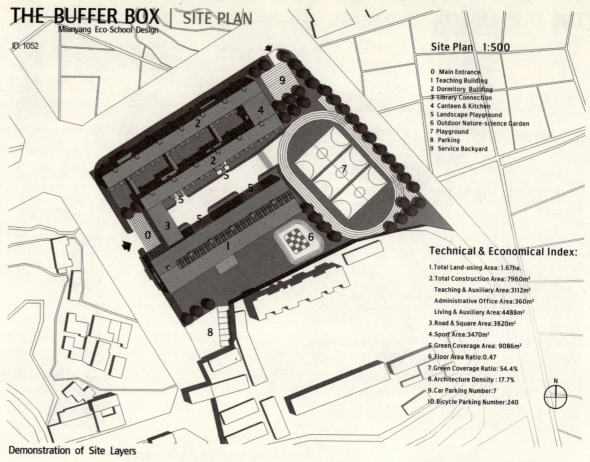

Site Plan 1:500

0 Main Entrance
1 Teaching Building
2 Dormitory Building
3 Library Connection
4 Canteen & Kitchen
5 Landscape Playground
6 Outdoor Nature-science Garden
7 Playground
8 Parking
9 Service Backyard

Technical & Economical Index:

1. Total Land-using Area: 1.67ha.
2. Total Construction Area: 7960m²
   - Teaching & Auxiliary Area: 3112m²
   - Administrative Office Area: 360m²
   - Living & Auxiliary Area: 4488m²
3. Road & Square Area: 3820m²
4. Sport Area: 3470m²
5. Green Coverage Area: 9086m²
6. Floor Area Ratio: 0.47
7. Green Coverage Ratio: 54.4%
8. Architecture Density: 17.7%
9. Car Parking Number: 7
10. Bicycle Parking Number: 240

Demonstration of Site Layers

04 / Structure Layer
After the earthquake in Sichuan 2008, structure of the building is emphasized. The main structure frame here is standard column-beam system. As the connection part, library has a steel bracing to keep stable.

03 / Plant Layer
Grass stretches on slopes in this layer, providing a safe and friendly surface for children. Selected plants which are out of diversity can adjust the micro-climate and reduce the noise from the road beside the site.

02 / Function Layer
Upon the basic district layer, the specific functions are settled in. Considering children's outside activities, besides the big playground, we design a landscape space for children to play and chat.

01 / District Layer
The basic functional division and basic route are arranged here. Fit to the site, we arrange the main route in northeast-southwest direction. Teaching part and living part are divided by the main route, at the same time, a bridge-library connects two parts together. Under the bridge-library, we make the entrance.

00 / Water Layer
Underground, the layer has a system to deal with collected water from rain and water circulation of the building. Coil pipes which use underground heat and water storage pool which store the heated water from solar water heating in summer holiday are also arranged in the layer.

Playground System

For children around 6 or 7 year old, playing outside the house is the most important part of education. Concrete steps built in grass provides a place for the little children to play with.

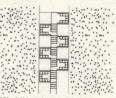

Children in higher grades are beginning to take activities such as communicating with each other. Seats on grass slope are places for children to get together and sit down, then do some chatting things.

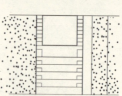

In primary school, an outside meultifunctional seat place is necessary because children have many activities such as meeting, performance, singsing......

Perspectives

Steps

Chatting Seats

Multifunctional Seats

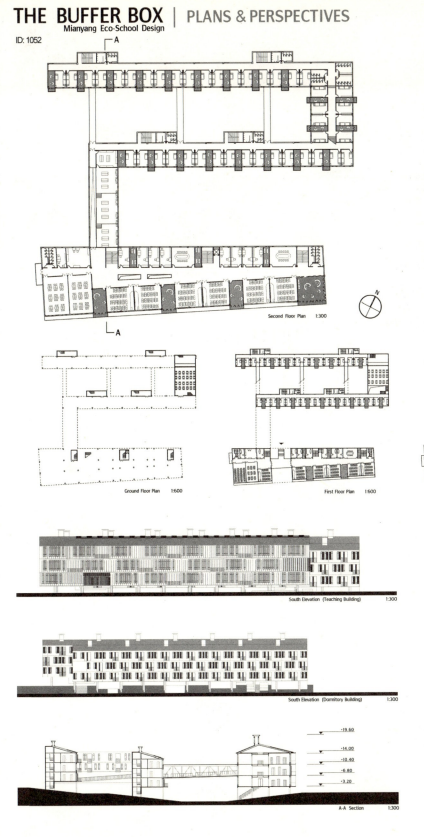

THE BUFFER BOX | PERSPECTIVE & UNITS
Mianyang Eco-School Design

Buffer Box - Space Changing

Except for the ecological use, Buffer Box also provide a space for students to play and communicate with each other. In dormitory, it can be a small public living room, while in classroom, it can be an additional area for teacher to organize particular teaching activities.

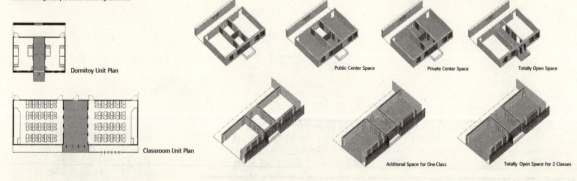

Buffer Box - Air Conditoning

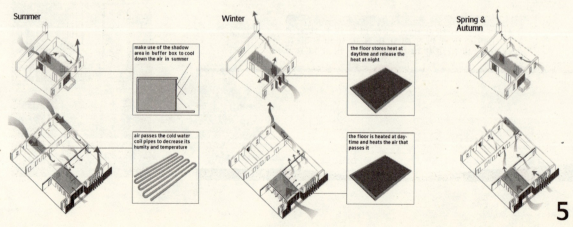

THE BUFFER BOX | TECTONICS, WATER SYSTEM & VENTILATION

Mianyang Eco-School Design
ID: 1052

TECTONICS & MATERIAL

The Buffer Box between Classrooms

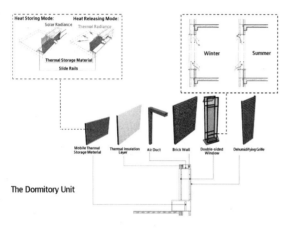

The Dormitory Unit

WATER SYSTEM

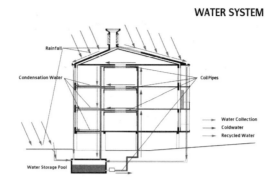

The Teaching Building

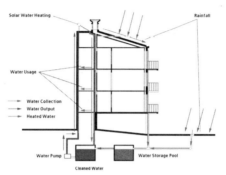

The Dormitory Building

VENTILATION

Summer

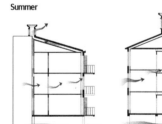
Dormitory Building

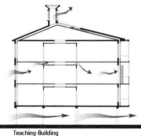
Teaching-Building

Winter

Dormitory Building

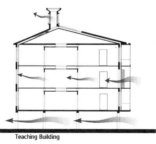
Teaching Building

Ventilation Simulation

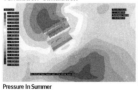
Pressure In Summer

Velocity In Summer

Pressure In Winter

Velocity In Winter

Analysis

The simulation results show that: under the climate of Mianyang, if a building is "thin" enough, wind can easily pass the building in a way we called through-drought, according to this, we overhead our building and make it as "thin" as possible. But the problem is that outside air enters the building with an uncomfortable temperature and humity, so we use a space which we called Buffer Box to make the air comfortable, then let the air enter the interior.

6

优秀奖
Honorable Mention Prize

项目名称：生态学校
　　　　　Eco-School
作　　者：王钰、刘伦
参赛单位：清华大学建筑学院建筑系

1078

MIANYANG
ECO–SCHOOL I

绵阳的气候最大的特点就是"湿"。一方面空气湿度大，一方面是降雨充沛。我们的设计即从此出发，针对绵阳气候采用了多种太阳能技术。

对于潮湿这一不利因素，我们采取了建筑布局通风、风压拔风、密闭风道送风等手段进行通风除湿。冬季由蓄热卵石加热来风，为教室提供暖风。

对于雨水这一资源，我们在供孩子活动的景观绿地下方形成潜流型人工湿地，收集过滤雨水，除作为景观用水，还作为太阳能集热器的水源。

绵阳不是太阳能充沛地区，我们采取了多种复合技术，并进行了简略的定量分析。

"Humid" is the most urging problem for the climate in Mianyang. First, humidity in air is heavy; on the other hand, Mianyang has abundant rainfall all over the year. Our design is based on it.

Faced with the disadvantage—humidification, we take steps including layout ventilation, wind pressure ventilation and ventilation by enclosed air passage etc. In winter we use pebble bed regenerator to heat up classrooms.

When it comes to rainfall, our solution is: jion constructed wetlands subsurface flow system with landscape greenbelt for children. Part of gathered rain used for landscape irrigation, part used for geothermal heating system.

Solar energy is not abundant in Mianyang, we have made quantitative analysis.

1078

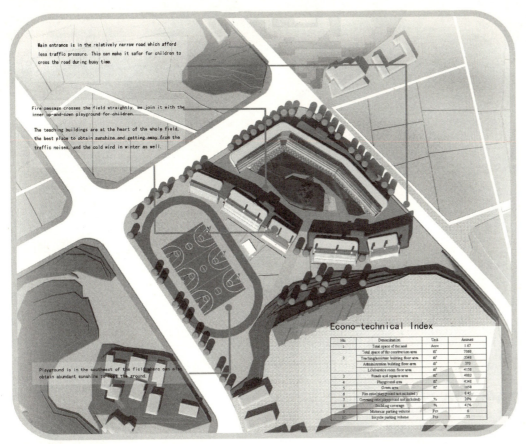

MIANYANG
ECO-SCHOOL

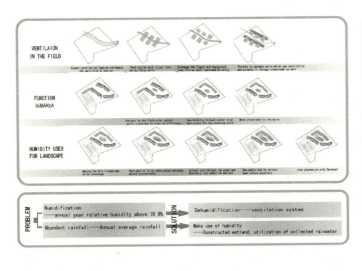

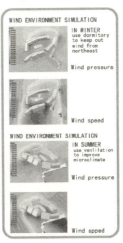

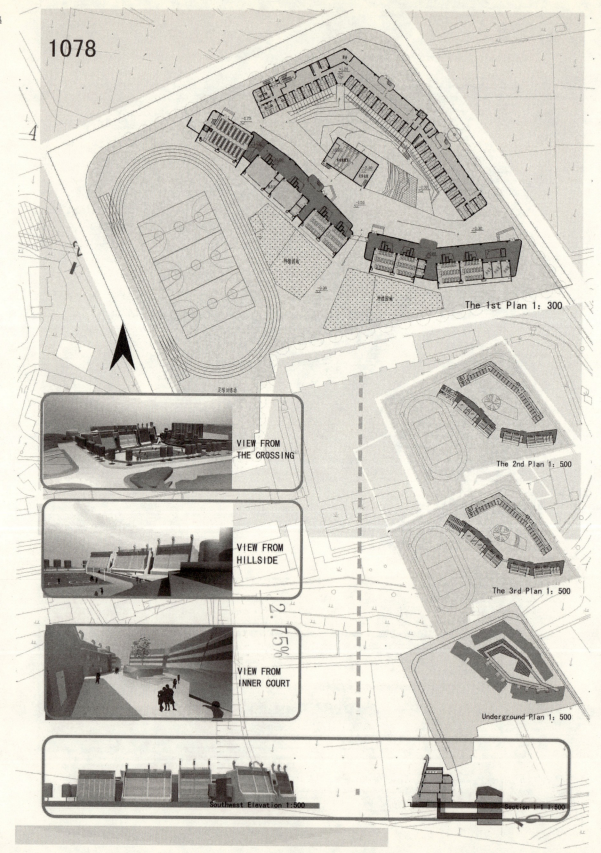

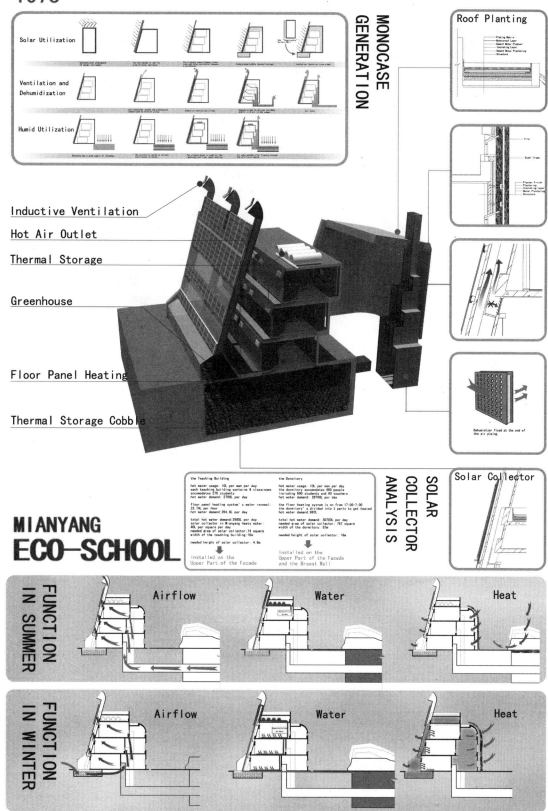

RURAL SUNSHINE PRIMARY SCHOOL IN MA ER KANG AREA

优秀奖
Honorable Mention Prize

项目名称：马尔康地区阳光小学
　　　　　Rural Sunshine Primary School In Ma Er Kang Area
作　者：朱健、欧克男
参赛单位：Bauhaus Dessau Foundation
　　　　　（德国德绍包豪斯研究院）

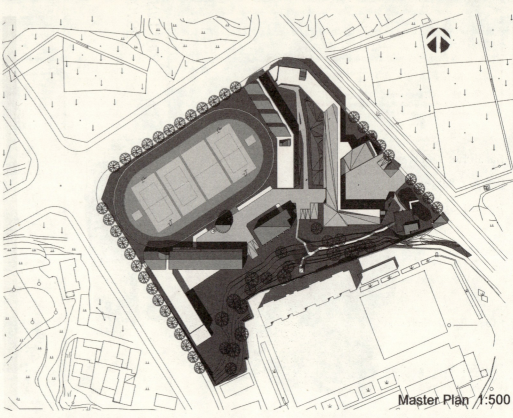

Master Plan 1:500

田野上的太阳房
大地剧烈震颤后，阳光依旧明媚。马尔康基地田里的庄稼郁郁葱葱，而孩子们等着，来年地里长出的将是太阳房子……

此方案力图设计一个布局合理而又有空间层次变化、形态错落有致、有生态内涵和人文精神的小学校园。根据当地气候等条件，综合而适配地利用太阳能等可持续建筑技术。在太阳能技术方面，除了利用建筑的基本构件——向阳大窗和"走廊式太阳房"以外，还特别设计了"大地太阳房"和"极小化太阳房"。"大地太阳房"灵感来自农业温室大棚，也旨在延续乡村的田野意象和生产活动。它创造了大面积附设太阳房的一个新形式，并且适用于任意朝向的建筑。"极小化太阳房"探索了空间极小化的太阳房形式，极大地节省占地和造价，兼有外墙美化和可分批加装等优点。

SOLAR HOUSES ON RURAL FIELD
After violent earthquake, sunshine is bright as it was. At the site of Ma Er Kang, rice field is expecting a coming harvest. Children are waiting, in the coming year, solar houses will grow on the field…

This proposal aims to design an interesting and properly structured school space system, a school in intrinsic ecological quality and good atmosphere of humanity. In the design, we utilized solar energy technology and other sustainable building technologies in a synthesized adaptable way. We not only made use of usual parts of a school building to utilize solar energy, such as window and corridor in sunshine, but also we especially designed 'Ground Solar House' and 'Minimum Solar House'. For 'Ground Solar House', the inspiration is from the agriculture greenhouse and also aim to continue the valuable agriculture field image and activity. It created a new type of large size solar house, and it can be adapted to a building of any orientation. 'Minimum Solar House' explored a space-minimized solar house for sun-facing wall. It is designed to extremely save land occupation and cost. It has also the aesthetics value for exterior wall and the advantage of installation in phases in terms of operability.

MAIN TECHNICAL AND ECONOMIC INDICATORS

Serial number	Item	Unit	Quantity
1	Total site area	hm²	1.67 (1.35*)
2	Total floor area of the building	m²	7935
	Floor area of teaching and accessory rooms	m²	2920
	Floor area of administration	m²	390
	Floor area of service	m²	4625
3	Road & square area	m²	1805
4	Sport field	m²	3688
5	Greening area	m²	4970
6	Floor area ratio (excluding sport field)	%	0.48 (0.59*)
7	Greening rate (excluding sport field)	%	30 (37*)
8	Building density	%	17 (21*)
9	Car parking spaces	car	7 (5 cars and 2 trucks)
10	Bike parking spaces	bike	74 (89m²)

* Total site area exclude town road area

RURAL SUNSHINE PRIMARY SCHOOL IN MA ER KANG AREA

LEGEND for FLOOR PLAN:

TEACHING AND ACCESSORY ROOMS
1. *Common classroom
4. *Nature class room
5. *Apparatus preparation room
6. *Computer room
7. *Computer accessorial room
11. *Skill activity room
14. *Sport equipment room

ADMINISTRATIVE ROOM
19. *Storage for general affairs
20. *Duty room

SERVICE ROOM
21. *Kitchen
22. *Canteen
23. *Boiling water room
24a. *Dormitory for boys (each bedroom with 4 school beds inside, for 8 people)
 #1 management room,
 #2 washrooms,
 #3 WC,
 #4 storeroom or clean tool room
25. *Dormitory for staffs (4 people a room, with WC, minimum kitchen counter and south balcony)
26. *Reception
27. Power distribution
BP. Bike parking
CP. Car parking

* must be established in task explanation.

ADDITIONAL DESIGN
29. 'Ground Solar House'
30. "Wisdom Hall", a landscape architecture reconstructed from the existing farmer's house.
31. Stage of the track and field ground (with photoelectric solar panel roof) & Flag-raising point.
32. "Solar Tower", a landscape architecture, also an additional stair for the adjoined teaching building, with photoelectric solar panel roof and wall paintings of stories of Sun.
33. Artificial marsh.

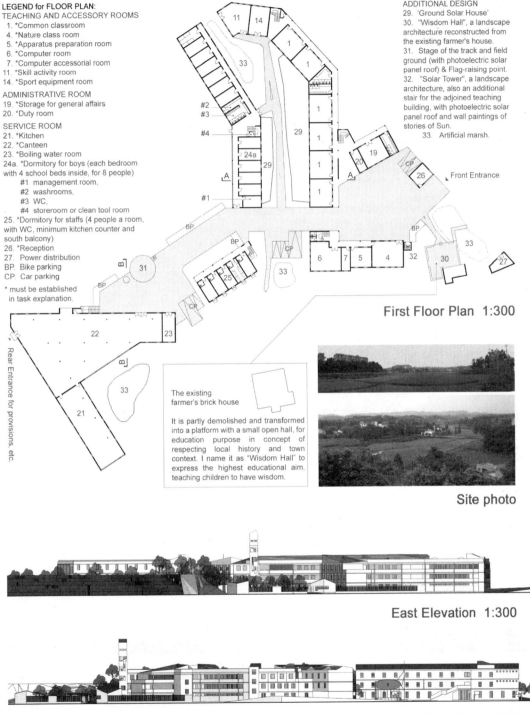

First Floor Plan 1:300

The existing farmer's brick house

It is partly demolished and transformed into a platform with a small open hall, for education purpose in concept of respecting local history and town context. I name it as "Wisdom Hall" to express the highest educational aim, teaching children to have wisdom.

Site photo

East Elevation 1:300

North Elevation 1:300

RURAL SUNSHINE PRIMARY SCHOOL IN MA ER KANG AREA

LEGEND for FLOOR PLAN:
TEACHING AND ACCESSORY ROOMS
1. *Common classroom
9. *Reading room for students
10. *Stack room
12. *Art room
13. *Art tool room

ADMINISTRATIVE ROOM
16. *Administration office
17. *Activity room for Young Pioneers
18. *Health office

SERVICE ROOM
24a. *Dormitory for boys (each bedroom with 4 bunk beds inside, for 8 persons)
  #1 management room, #2 washrooms, #3 WC, #4 storeroom or clean tool room
24b. *Dormitory for girls (same as boys')
25. *Dormitory for staffs (each bedroom has 4 high raised single beds with desks and wardrobes under beds, for 4 persons)
29. 'Ground Solar House'
31. Stage of the track and field ground
32. "Solar Tower"
34. 'Minimum Solar House' panel
35. 'Corridor Solar House'

* must be established in task explanation.

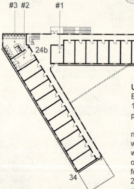

Utilization of active solar energy technology:
1. Utilization of solar water heater. It is already a very developed solar energy technology with high performance-cost ratio. It is designed to be on part of roofs of the three dormitory buildings, for hot water and room heating.
2. Utilization of photoelectric solar panels. It is designed for roofs of the stage of the track and field ground and the 'Solar Tower', No.31 & 32. Part of the three dormitory building roofs will be also covered with photoelectric solar panels, at proportion ballenced with utilization of solar water heater.

Second Floor Plan 1:300

Utilization of passive solar energy technology:
Because of the site limitation and the room sunshine requirement, we developed some new design ideas.
1. 'Ground Solar House'. Solar houses in agricultural greenhouse style on the ground and on the roof, see No.29 parts in Second Floor Plan, , and also see A-A Section.
  Usually a solar house is designed as a transparent space covering a south wall area to catch sunlight. Here this new type of solar house is a space covering a sun shined ground area to catch sunlight. It is very fit to buildings with north-south linear form. This idea is from the context of agriculture land with greenhouses. Here it is integrated with rural area school buildings not only for building rooms warmed by it, but also for planting proper vegetables or flowers as a productive land as before to support the school, and it is a part of exterior court for natural science for education.
2. 'Minimum Solar House', glass panel for exterior wall, see page 4, the detail drawing and B-B section. It is a glass panel with its perimeter air-tightly attached onto a wall area and leaves just 4-10 cm interspace for air circulation. It is designed to cover most parts of the south, south-west and south-east walls of the school buildings, see Second Floor Plan, red lines marked walls.
3. 'Corridor Solar House', the first to third floor west corridors of the dormitory for boys are designed as solar houses, see A-A section on next page.

West Elevation 1:300

(for this dawing, south hill is hidden but the path and kiosk are still there) **South Elevation 1:300**

RURAL SUNSHINE PRIMARY SCHOOL IN MA ER KANG AREA

Adoption of other technology:
1. Utilize wind power to fan warm air into lower rooms from the roof 'Ground Solar House' and to circulate heated water from roof heaters to lower heat radiator. See A-A & B-B Sections.
2. Design part of the green area as artificial marsh for school wastewater, stormwater runoff, and as habitat for wildlife. In fact it is a kind of keeping the land similar to it was, the rice field. See the first floor plan, No. 33.

Operability of the technology:
1. Low additional budget technology, the west corridors being designed as solar houses, is firstly adopted.
2. Some technologies are adopted or developed for new designs because they are relatively of not too high cost in terms of additional budget, and they can be added in phases, and some of them have not only the value of solar energy. The 'Minimum Solar House' glass panel has also aesthetics and waterproof values for exterior wall. The 'Ground Solar Houses' in agriculture greenhouse style also has the value for agriculture and education. Both of them and the roof solar water heaters can be installed in phases.
3. Relatively high cost advanced solar energy technologies are encouraged but it need to be designed according to budget, such as photo-electric solar panel. And some are abnegated, such as underground water compartment for heat storage.

LEGEND for FLOOR PLAN:
TEACHING AND ACCESSORY ROOMS
1. *Common classroom
2. *Music classroom
3. *Instrument room
8. *Multi-function room
15. *Office room

ADMINISTRATIVE ROOM
16. *Administration office

SERVICE ROOM
24a. *Dormitory for boys
24b. *Dormitory for girls
 #4 storeroom or clean tool room
25. *Dormitory for staffs

* must be established in task explanation.

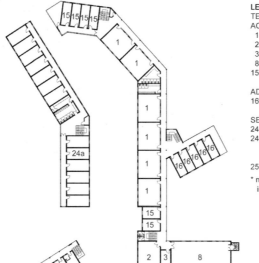

Third Floor Plan 1:300

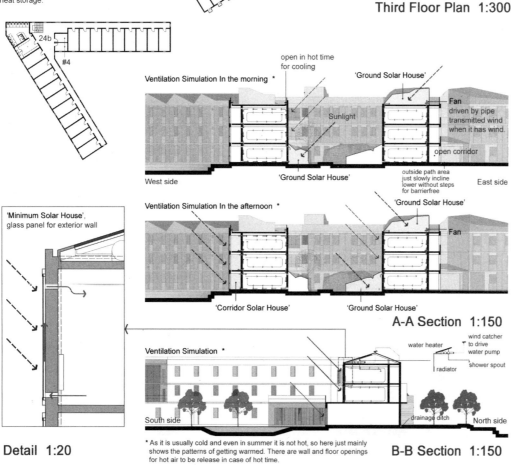

Detail 1:20

A-A Section 1:150

B-B Section 1:150

* As it is usually cold and even in summer it is not hot, so here just mainly shows the patterns of getting warmed. There are wall and floor openings for hot air to be release in case of hot time.

RURAL SUNSHINE PRIMARY SCHOOL IN MA ER KANG AREA

Bird View From East

Perspective of the front plaza

Perspective from the 'Wisdom Hall' to hillside green area

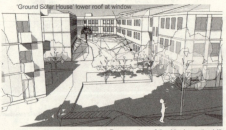

'Ground Solar House' lower roof at window
Perspective of the kiosk on the hill

Perspective from the 'Ground solar House' area to the kiosk

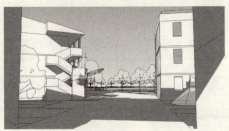

Perspective from the teaching building gate

Perspective from the sports field to the Canteen and girls' dormitory

RURAL SUNSHINE PRIMARY SCHOOL IN MA ER KANG AREA

LEGEND:
1. Slope roof for solar water heater and may also for photoelectric solar panel in future.
2. 'Ground solar House' on roof.
3. Flat roof part for life use, roof planting and also surfaces for future installation of photoelectric solar panel, solar water heater, 'Ground solar House' on roof or other equipment.
4. Basketball and volleyball field.
5. Small football field.
6. Volleyball and small football field.

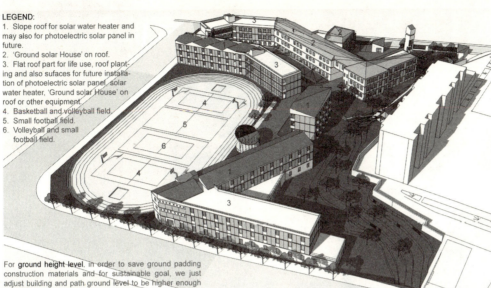

For **ground height level**, in order to save ground padding construction materials and for sustainable goal, we just adjust building and path ground level to be higher enough for waterproof and keep the green area and sports field part to be on the same level as it was as possible. It is also fit to the design of artificial marshes in terms of ground level.

Bird View From Southwest

Perspective of the front entrance

Perspective of the 'Wisdom Hall' and an artificial marsh

Birdview of the front plaza area

Birdview of the main green area along the hillside

Photoelectric solar panel
Perspective of the Stage of sports field

'Minimum Solar House' panel
Perspective of the hillside and beside buildings

175

优秀奖
Honorable Mention Prize

项目名称：马尔康之花
Blossom at Ma Er Kang
作　者：Gádor Luque. Sara Rojo. Victor Quirós
参赛单位：Gádor Luque. Sara Rojo. Victor Quirós

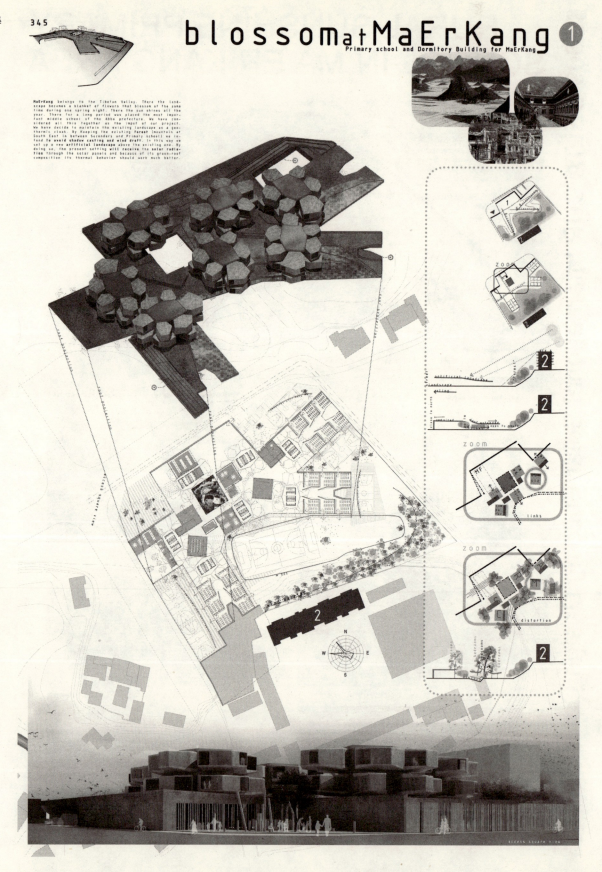

blossom at MaErKang ②
Primary school and Dormitory Building for MaErKang

IN THE SCHOOL BUILDING The interior space is a body where three arms that contains six common classrooms each, hug the nature and the sports areas. The trunk is perforated by the patio (conceived as a Chinese Garden) that generates a natural ventilation and introduce vegetation inside the building. Around this poetic space the special classrooms are placed according to different requests as they were organs.

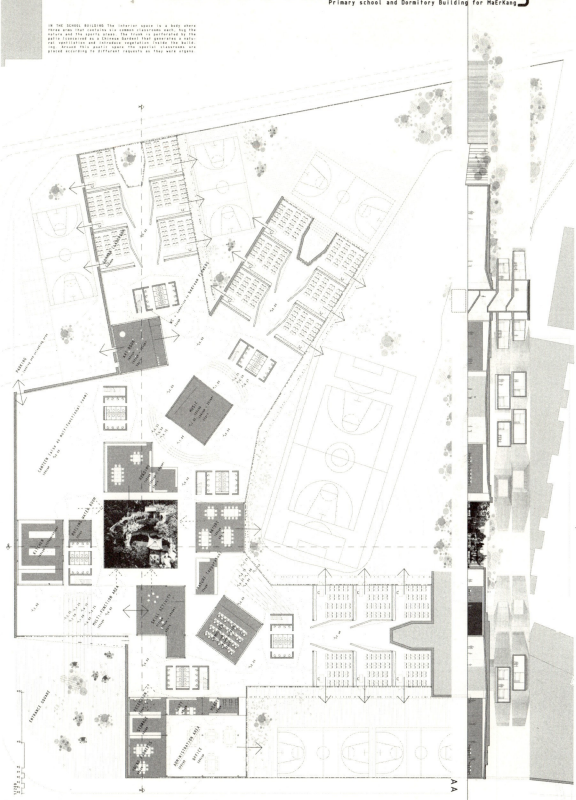

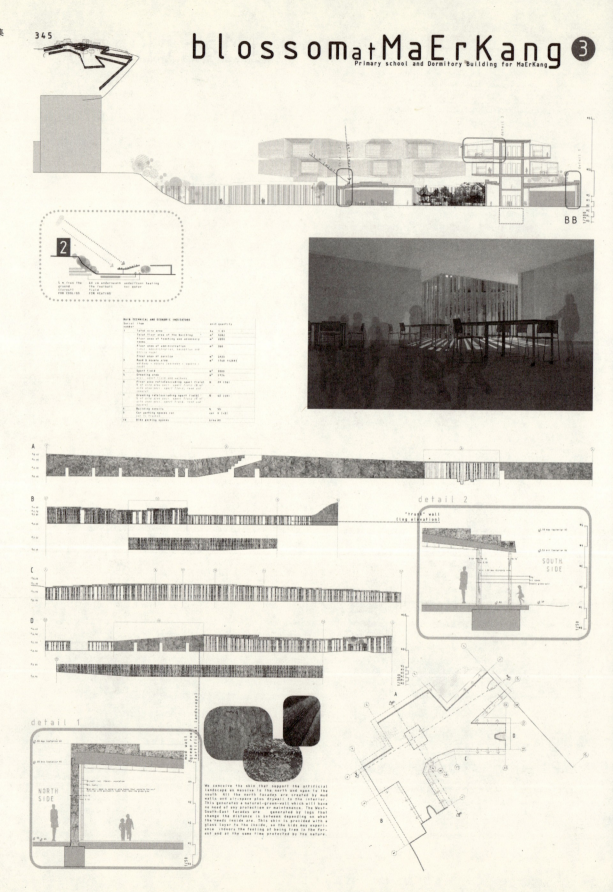

blossom at MaErKang
Primary school and Dormitory Building for MaErKang

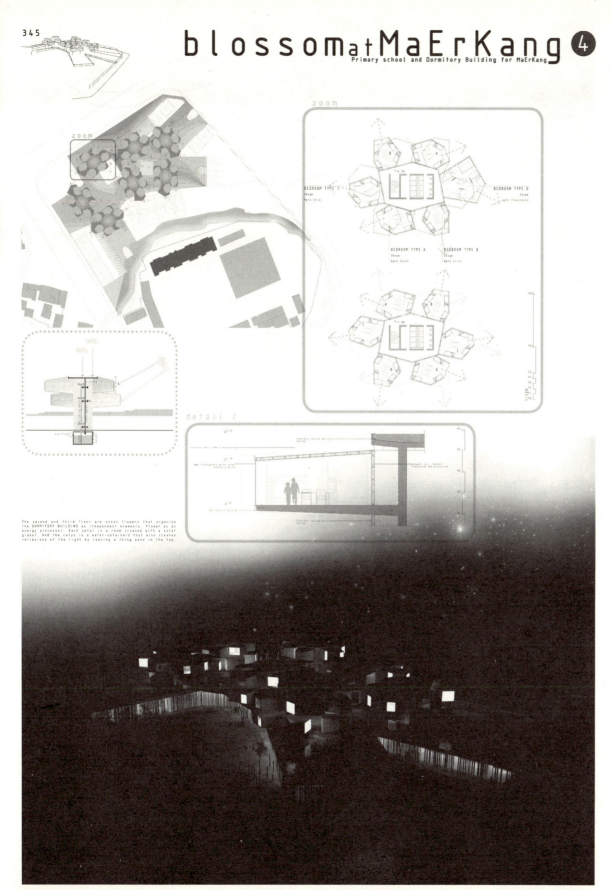

优秀奖
Honorable Mention Prize

项目名称：370号作品
　　　　　No.370
作　　者：汪海涛、张妍、姚亦梅、谭祈燕
参赛单位：西南交通大学建筑学院

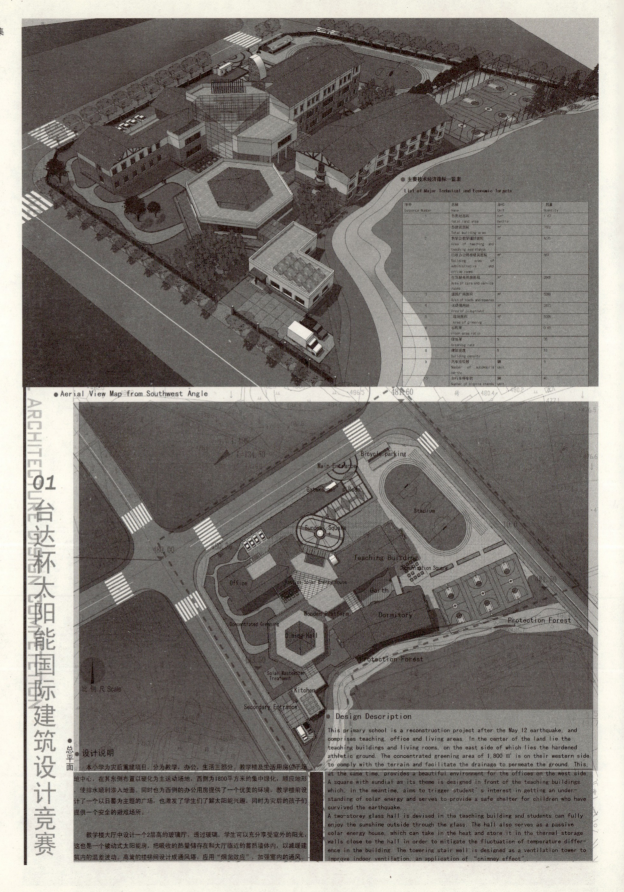

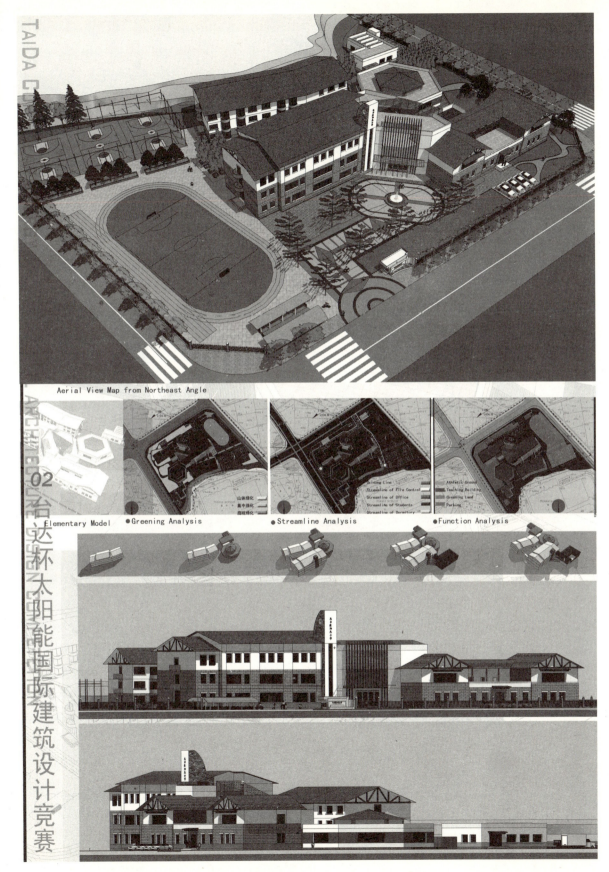

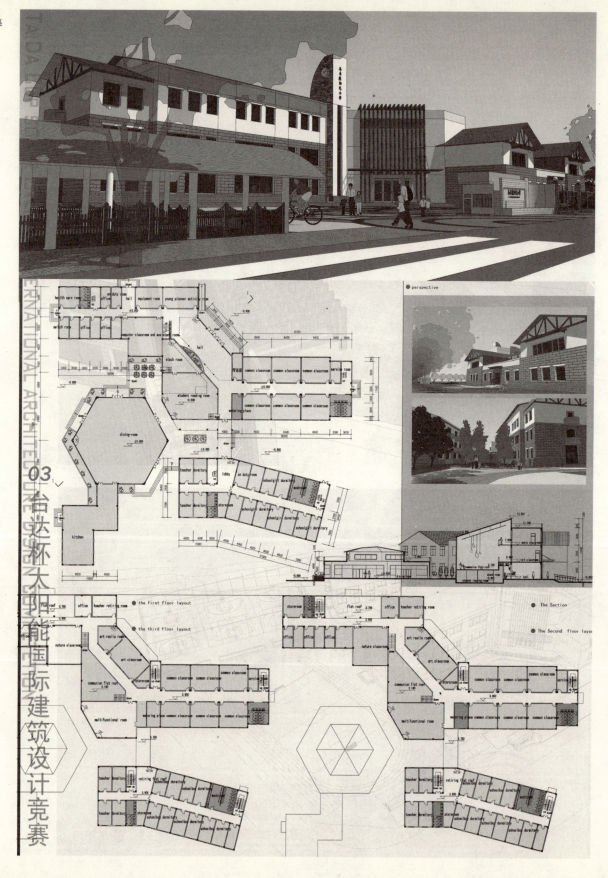

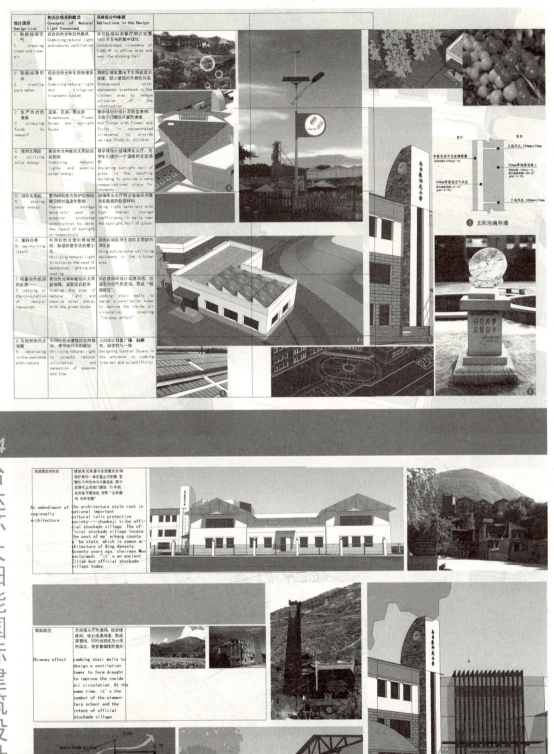

马尔康阳光小学设计
一半是记忆，一半是未来

优秀奖
Honorable Mention Prize

项目名称：一半是记忆，一半是未来
　　　　　Half is memory, half is future
作　　者：张超、侯薇、刘伟、朱江涛、
　　　　　孟庆山、唐一萌、李莺
参赛单位：河北工业大学

设计说明：

阳光小学位于马尔康农村地区，整体朝向为西南向，在保证日照的条件下，顺应当地建筑布局，尊重其历史形态。

小学建筑由教学区和生活区两部分组成，动静分离。教学区由依托于一个基底平台的三个单体围合而成，在增强联系性的同时，提供更多的疏散途径。

阳光小学采用被动、主动式太阳能技术相结合的方式，南墙应用特朗勃墙体系，巧妙设置太阳房，屋顶大面积太阳能集热器，辅以低温热水地板辐射采暖，实现生活及采暖热水的供应。

建筑材料以再生材料为主，且采取措施以减少石化材料的用量，以助于生态环境的保护。

Idee

Set in the countryside, in a condition of enough sunshine, conforms to the local construction layout, in respect of local historic form.

The building is composed of school districts and utility area to separate the activity zone. School districts is surrounded by three single building based on the same terrace, providing more evacuation route and strengthening accessibility simultaneously.

The Sunshine Primary School combines passive and initiative solar energy utilization. South wall uses Trombe Wall, applying solar house ingeniously. The roof is laied solar collector widly, adding low temperature hot water floor radiant heating as a supplement, to offer demestic water use and heating.

The architectural material give priority to the recyclable materials and take measures to minimize the use of petrochemical materials to protect the environment.

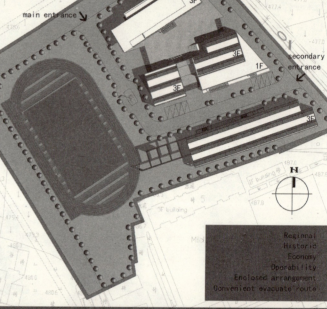
Site Plan 1:500

Reginnal
Historic
Economy
Oporability
Enclosed arrangement
Convenient evacuate route

Generation of the volumes
Terrace —— Surrounding —— separation of the activity zone

We won't lost that memory, nor will we lost our hope for the future. Because of the memories, there are the buildings; because of the children, there are dreams of the future.

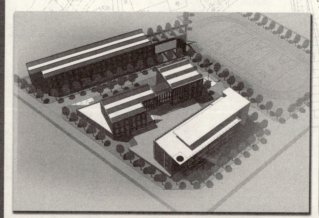

Number	Item	Units	Quantity
1	Total land area	m²	16708
2	Total construction area	m²	7021
	Teaching room	m²	2653
	Administrative office	m²	351
	Life of service space	m²	4017
3	Road and Plaza land	m²	1800
4	Schoolyard	m²	3504
5	Green area	m²	4709
6	Floor area ratio		0.53
7	Greening rate	%	35.66
8	Building density	%	21.04
9	Car parking spaces		6
10	Bike parking		300

ECONOMIC INDICATORS

2009 Delta Cup-International Solar Building Design Competition

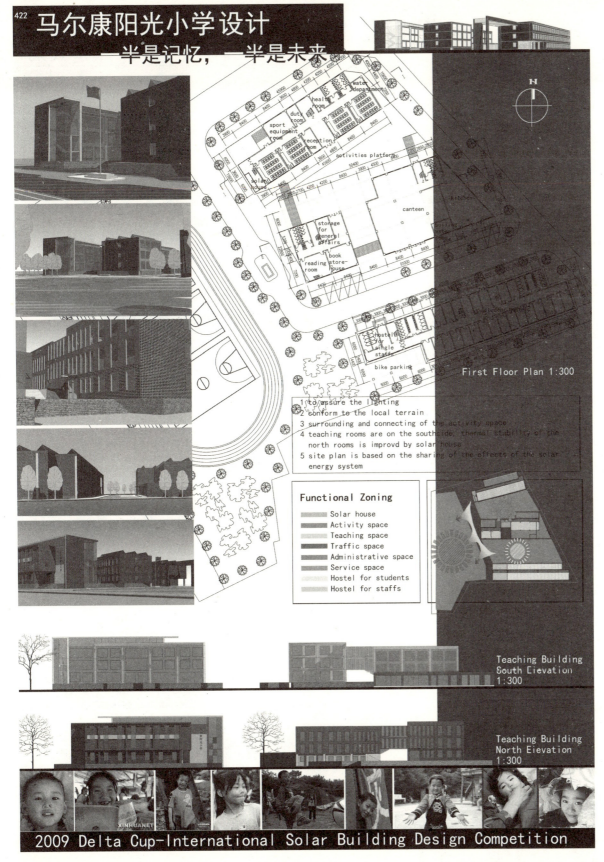

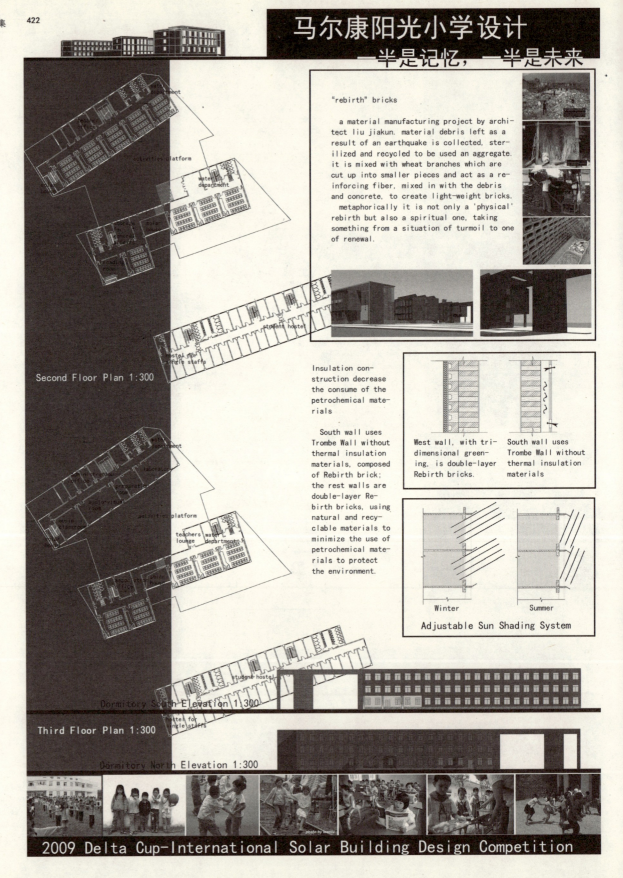

马尔康阳光小学设计
一半是记忆，一半是未来

Winter day

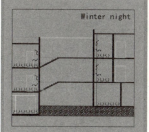
Winter night

Summer day

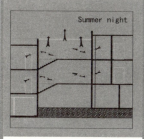
Summer night

Heat circulation Schematic illustration

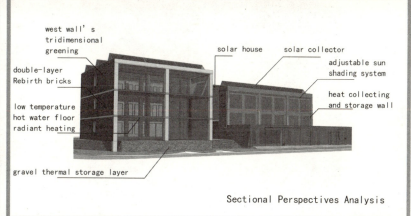

west wall's tridimensional greening
double-layer Rebirth bricks
low temperature hot water floor radiant heating
gravel thermal storage layer
solar house
solar collector
adjustable sun shading system
heat collecting and storage wall

Sectional Perspectives Analysis

Energy untilization methods:
Solar collect system: to provide 100% of domestic hot water and 40% of low temperature hot water floor radiant heating.
Supplementary system: hot pump integrated in the solar collector system, with solar as the evaporator, provides low temperature heating about 10 to 20 centi degree.

Solar house and architectural materials

The traffic space is used as solar house, which is convenient to give heat and natural ventilation. Large-area lucid interspace, together with west wall's tridimensional greening provides a better view from the outside and inside of the buildings. Inosculation of several materials manifests enormous vitality.

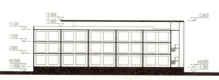
I-I Section 1:300

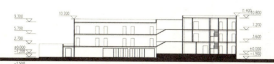
II-II Section 1:300

2009 Delta Cup-International Solar Building Design Competition

优秀奖
Honorable Mention Prize

项目名称：阳光 & 梦想
　　　　　Sunshine & Dream
作　者：张蕊、邓静静、张立杰
参赛单位：滨州市规划设计研究院

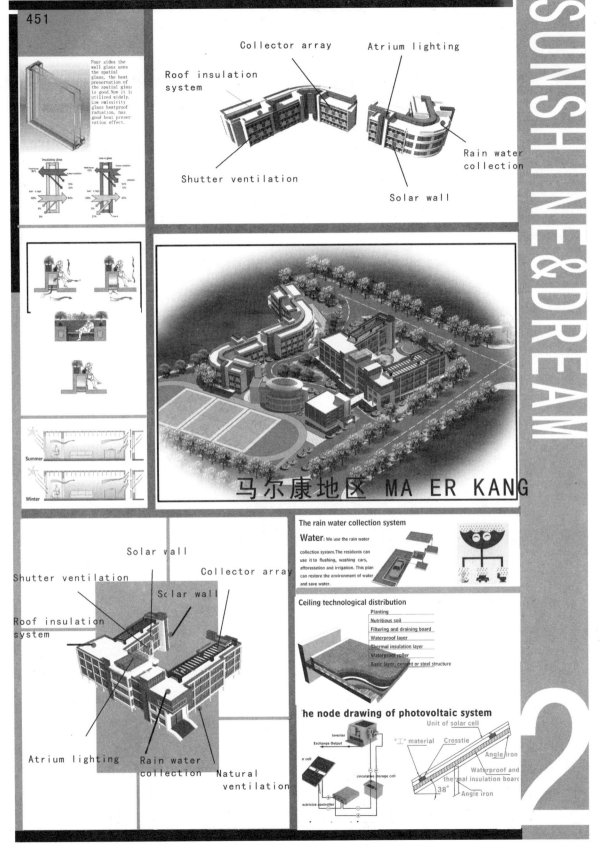

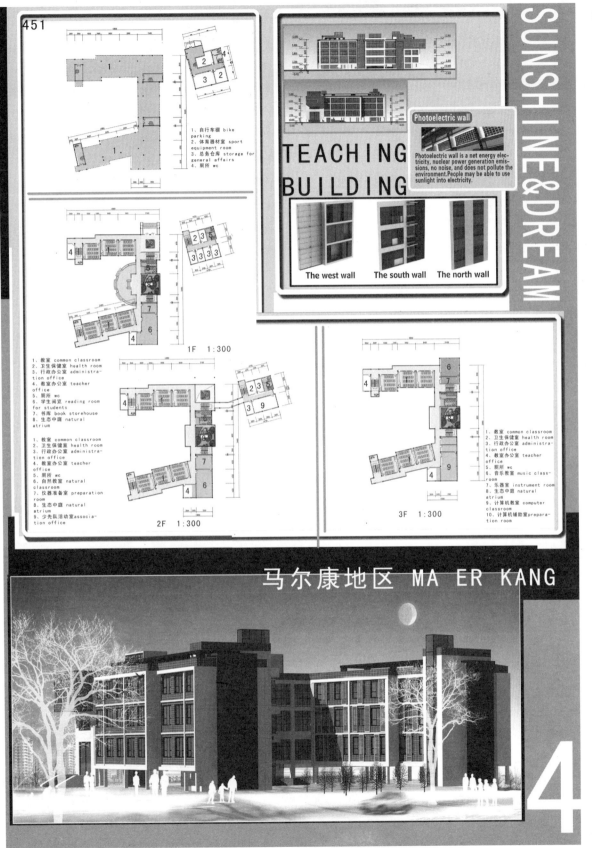

优秀奖
Honorable Mention Prize

项目名称：阳光面对面
　　　　　Face To Sunshine
作　　者：王鹏、于瑶
参赛单位：北京市建筑设计研究院

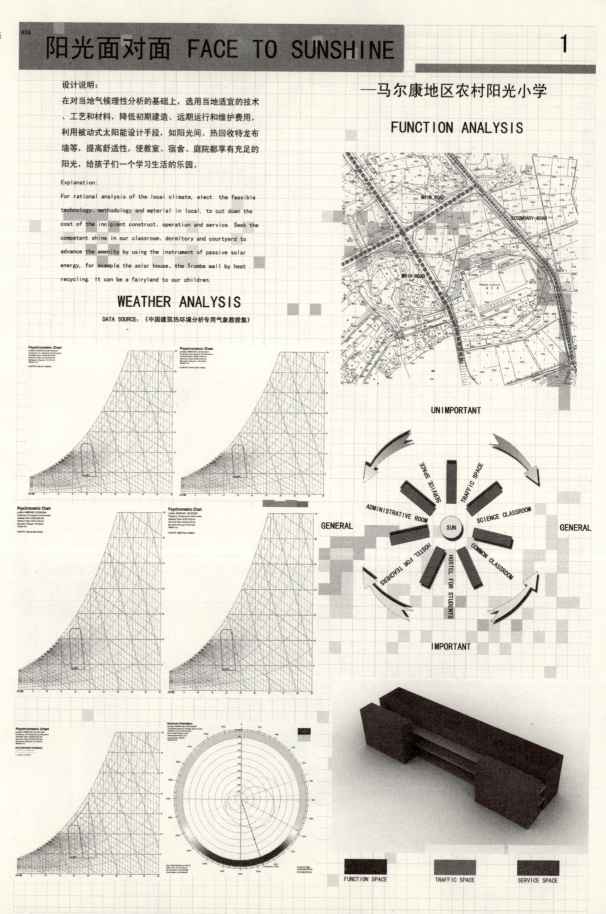

阳光面对面 FACE TO SUNSHINE

—马尔康地区农村阳光小学

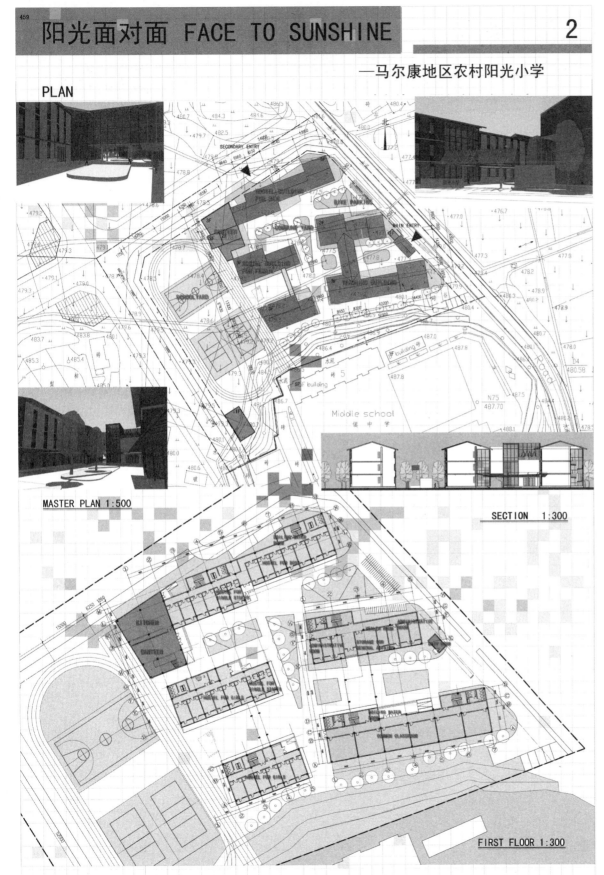

阳光面对面 FACE TO SUNSHINE

—马尔康地区农村阳光小学

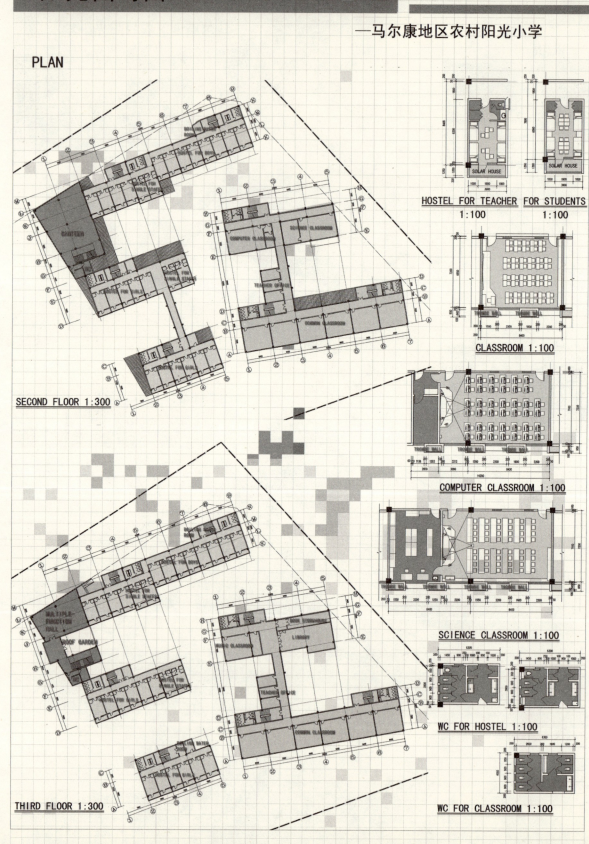

阳光面对面 FACE TO SUNSHINE

—马尔康地区农村阳光小学

SOLAR THERMAL FOR BATHROOM

- tile
- air
- thermal insulation
- concrete

ROOF DETAIL 1:20

- tile
- air
- thermal insulation
- concrete

- compact soil block
- air
- thermal insulation
- concrete

WALL DETAIL 1:20

VENTILATIONG BY SIDE
CONTROLLABLE SWITH
CONTROLLABLE SHUTTER
HEAT RECYCLING DEVICE
CONTROLLABLE SHUTTER
CONTROLLABLE SHUTTER
VERTICAL AXIS WIND TURBINE

TROMBE WALL BY HEAT RECYCLING DETAIL 1:20

序号	名称	单位	数量
1	总用地面积	hm²	1.67
2	总建筑面积	m²	7974.00
	教学及教学辅助用房建筑面积	m²	3069.00
	行政办公用房建筑面积	m²	372.00
	生活服务用房建筑面积	m²	4533.00
3	道路广场面积	m²	7298.00
4	运动场地面积	m²	3488.00
5	绿地面积	m²	3256.00
6	容积率（不含运动场地）		0.60
7	绿地率（不含运动场地）	%	19%
8	建筑密度	%	16%
9	汽车泊位数	辆	6.00
10	自行车停车数	辆	66.00

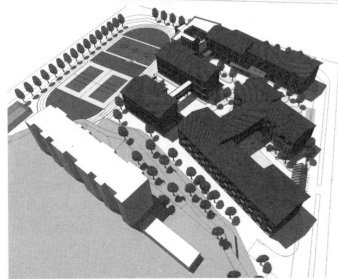

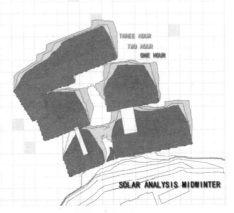

THREE HOUR
TWO HOUR
ONE HOUR

SOLAR ANALYSIS MIDWINTER

优秀奖
Honorable Mention Prize

项目名称：阳光温室
　　　　　Sunlight Greenhouse
作　者：刘婷婷、刘烨、王鑫
参赛单位：天津大学建筑学院、
　　　　　清华大学建筑学院

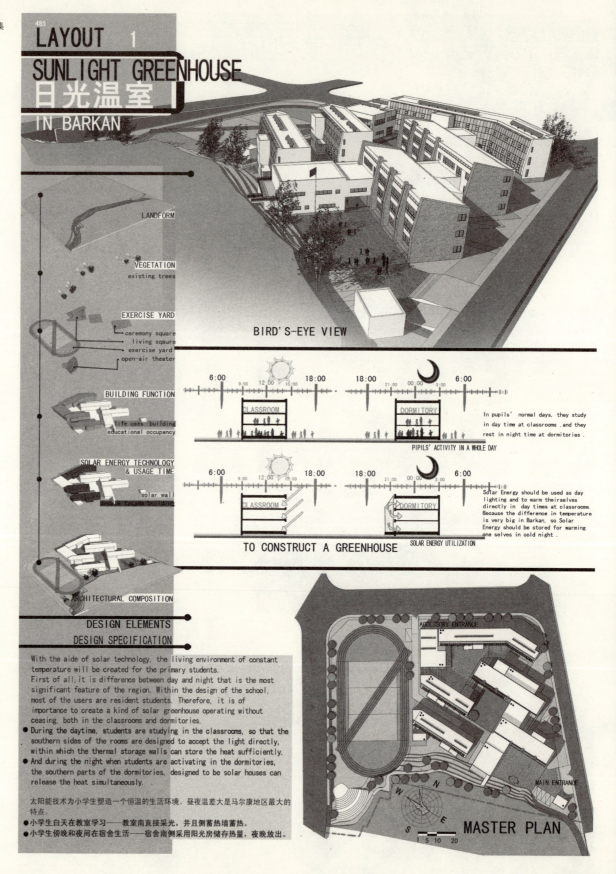

LAYOUT 1
SUNLIGHT GREENHOUSE
日光温室
IN BARKAN

BIRD'S-EYE VIEW

- LANDFORM
- VEGETATION — existing trees
- EXERCISE YARD — ceremony square / living square / exercise yard / open-air theater
- BUILDING FUNCTION — life uses building / educational occupancy
- SOLAR ENERGY TECHNOLOGY & USAGE TIME — solar wall / double facade building
- ARCHITECTURAL COMPOSITION

PIPILS' ACTIVITY IN A WHOLE DAY

In pupils' normal days, they study in day time at classrooms, and they rest in night time at dormitories.

TO CONSTRUCT A GREENHOUSE — SOLAR ENERGY UTILIZATION

Solar Energy should be used as day lighting and to warm theirselves directly in day times at classrooms. Because the difference in temperature is very big in Barkan, so Solar Energy should be stored for warming one selves in cold night.

DESIGN ELEMENTS
DESIGN SPECIFICATION

With the aide of solar technology, the living environment of constant temperature will be created for the primary students.
First of all, it is difference between day and night that is the most significant feature of the region. Within the design of the school, most of the users are resident students. Therefore, it is of importance to create a kind of solar greenhouse operating without ceasing, both in the classrooms and dormitories.
- During the daytime, students are studying in the classrooms, so that the southern sides of the rooms are designed to accept the light directly, within which the thermal storage walls can store the heat sufficiently.
- And during the night when students are activating in the dormitories, the southern parts of the dormitories, designed to be solar houses can release the heat simultaneously.

太阳能技术为小学生塑造一个恒温的生活环境。昼夜温差大是马尔康地区最大的特点。
- 小学生白天在教室学习——教室南直接采光，并且侧蓄热墙蓄热。
- 小学生傍晚和夜间在宿舍生活——宿舍南侧采用阳光房储存热量，夜晚放出。

MASTER PLAN

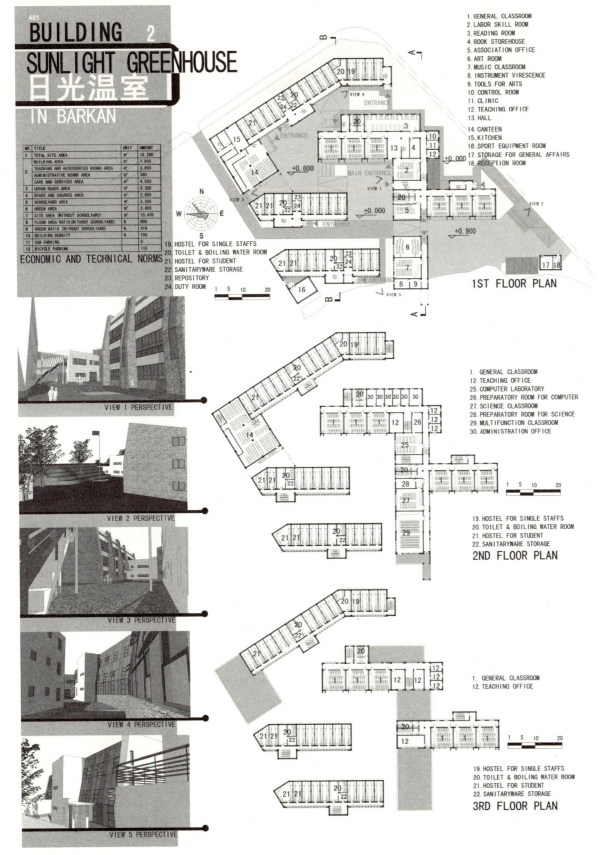

EASY TO BE BUILT 4
SUNLIGHT GREENHOUSE
日光温室
IN BARKAN

马尔康当地建筑很有特色，被称作"木与石的乐章"，在设计中，考虑到建筑的易建造性，在任务书上要求的框架结构体系中，自由的加入了当地的木与石，有效的降低建筑成本，也使新建筑与原有文脉呼应。

Barkam local building is very unique, and has been called "the movement of wood and stone". In this design, we take easy construction into account. In the mission the book frame structure of system is required, we join local wood and stone freely and effectively reduce the construction cost, but also new building echoes the original context.

MORDERN STYLE =

WOOD: Wood are one of local materials, they are easy to get and they are cheap. Nowadays, local people still use it to

STONE: Stone are one of local materials, they are easy to get. Stone is one of the most characterized materials in Barkan.

GLASS: Glass is one of common materials which are used in solar energy utilizations, and they are easy to get and cheap.

WOOD + STONE = BUILDING ↓ TRADITIONALSTYLE

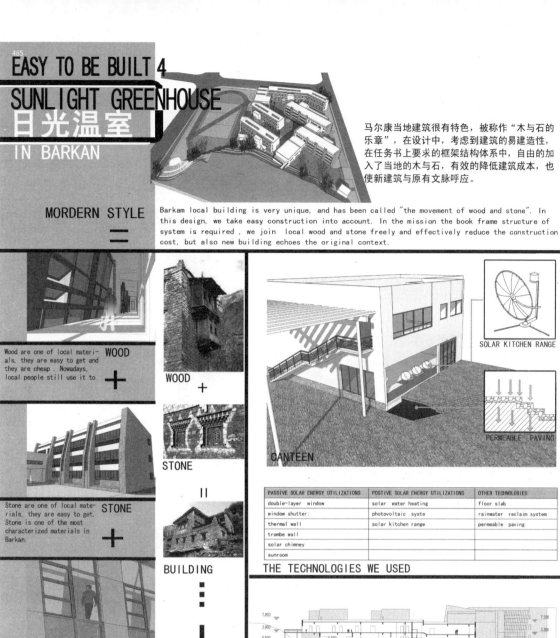

CANTEEN
SOLAR KITCHEN RANGE
PERMEABLE PAVING

PASSIVE SOLAR ENERGY UTILIZATIONS	POSTIVE SOLAR ENERGY UTILIZATIONS	OTHER TECHNOLOGIES
double-layer window	solar water heating	floor slab
window shutter	photovoltaic syste	rainwater reclaim system
thermal wall	solar kitchen range	permeable paving
trombe wall		
solar chimney		
sunroom		

THE TECHNOLOGIES WE USED

SECTION A-A

SECTION B-B

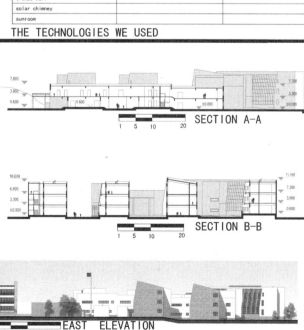

SOUTH ELEVATION   EAST ELEVATION

优秀奖
Honorable Mention Prize

项目名称：阳光猎人
　　　　　Sunshine Hunter

作　者：屈天鸣、贾子健、张仕寅

参赛单位：华中科技大学建筑与城市规划学院
　　　　　绿色建筑研究中心

SUNSHINE HUNTER — 526 — BARKAM SOLAR PRIMARY SCHOOL DESIGN COMPETETION

The average temperatures for the county of Barkam go from -0.5°C for the coldest month to +16.2°C for the hottest month, with an annual average of +8.7°C, and the Pluviometrie is there of 753 mm.

BARKAM
Location: 31.5° N   102.2° E
Altitude: 2866 m
Climate: Qinghai-Tibet Plateau Cold

REGION

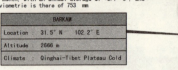

中国气候类型

CLIMATE ANALYSIS

Generally speaking, it is cold in winter and cool in summer in Barkam. Inhabitants use the schist to build the thick wall, it is very efficacious to reduce the thermal conductivity.

Small windows are used in order to keep warm. But meanwhile, less light and heat can get into the house through the windows.

LOCAL STYLE

LEARN FROM VERNACULAR

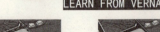

STUDENTS' SCHEDULE

0:00-7:30 dormitory | 8:00-12:00 classroom | 13:00-14:00 dormitory | 16:30-17:30 playground | 18:30-21:30 dormitory
7:30-8:00 canteen | 12:00-13:00 canteen | 14:00-16:30 classroom | 17:30-18:30 canteen | 21:30-24:00 dormitory

major road / secondary road　　possible position for main entrance

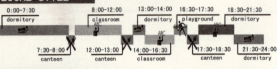

9:00 10:00 11:00 12:00

So, in order to get heat mostly and early, classrooms lay in the south-east part.

Morning, the sun shines the south-east part of the site. Meanwhile, the students are using classrooms.

13:00 14:00 15:00 16:00

So in order to storage heat in the afternoon and supply heat at night, dormitories lay in the south-west part.

TIME AND SPACE

Afternoon, the sun shines the south-west part of the site. Meanwhile, the dormitories are storing heat.

设计说明：
■ 本次小学设计中主要解决冬季教室和宿舍的采暖储热问题。
■ 作息时间与空间：根据小学的作息时间和冬至日太阳运行轨迹，确定小学的总体布局。教室主要利用东南向的光照，宿舍主要利用西南向的光照。
■ 传统民居的启示：以厚重的墙体和活动风门满足保温和通风要求，并采用当地石材和墙体砌筑技术。
■ 被动式太阳能技术支撑：教室南向以可调节的双层玻璃幕墙采光集热。宿舍运用温室和蓄热墙形成的图洛姆墙体系进行集热，并结合文丘里效应和烟囱效应进行通风。

■ The main point during this primary design is how to get the heat and keep rooms warm in winter.
■ STUDENTS' SCHEDULE AND SPACE: The Site plan design is according to the students' schedule and the sun track. Classrooms settle in southeast in order to get heat in the early morning. On the contrary, the light from southwest heats the dormitories.
■ LEARNING FROM VERNACULAR: Using the thick wall and wind gate in north meet the insulation and ventilation requirements;
Using local appropriate technical and material build the wall.
■ PASSIVE SOLAR ENERGY TECHNICAL SUPPORT: The double-glazed windows and Trombe wall in south of classrooms meet the requirement of light collection. Greenhouse and thermal storage wall make up the surface of dormitory and form collector wall system, combined with Venturi effect and ventilation chimney effect.

DESIGN ILLUSTRATION

No.	Name	Units	Amount
1	total land area	Ha	1.67
2	total construction area	m²	7866
	teaching and teaching aids service space	m²	3117
	adminstrative room	m²	498
	services space	m²	4251
3	road and plaza	m²	5740
4	sports ground	m²	3329
5	green	m²	4033
6	plot ratio (non-sport ground)		0.59
7	ratio of green space (non-sport ground)	%	30.2
8	building density	%	21.5
9	number of car parking		5
10	number of bicycle parking		70

TECHNICAL AND ECONOMIC INDEXE

SITE PLAN 1:500

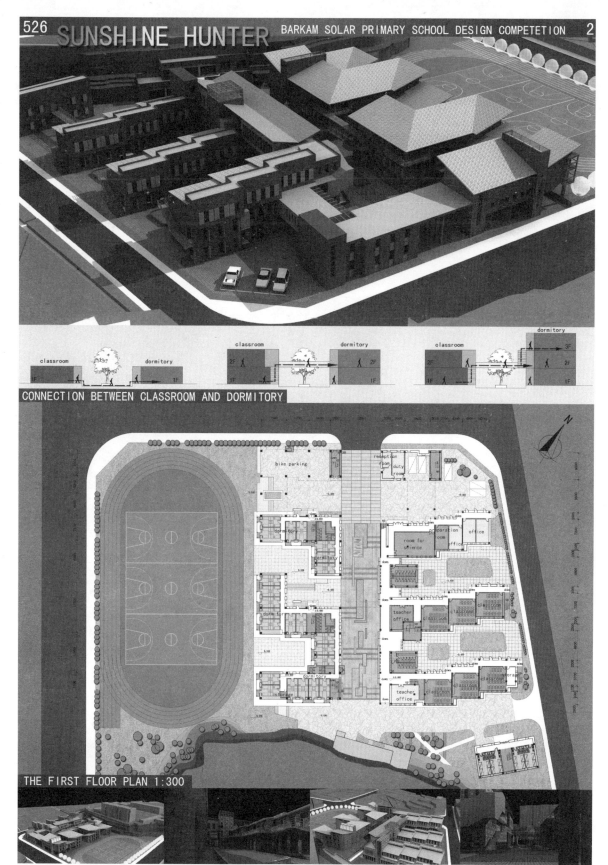

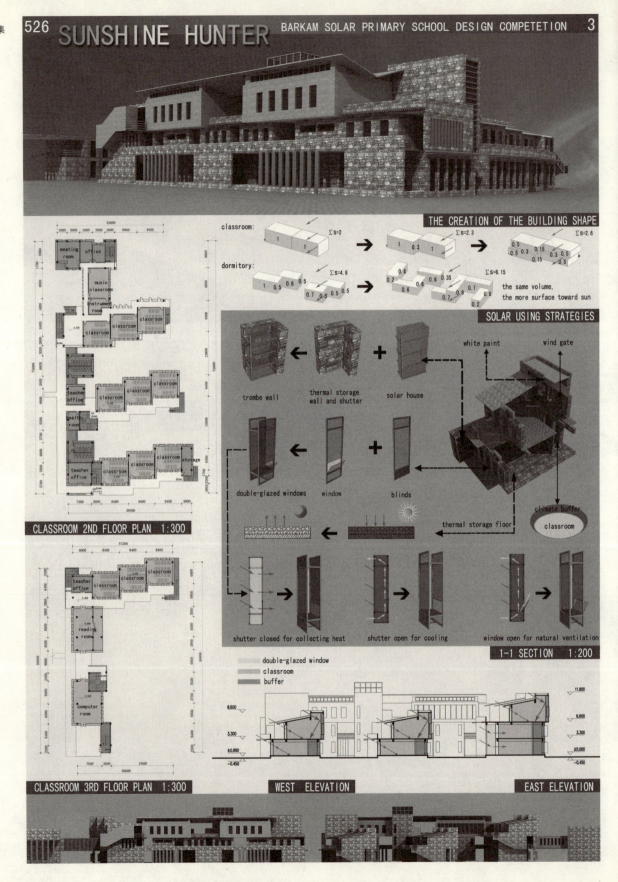

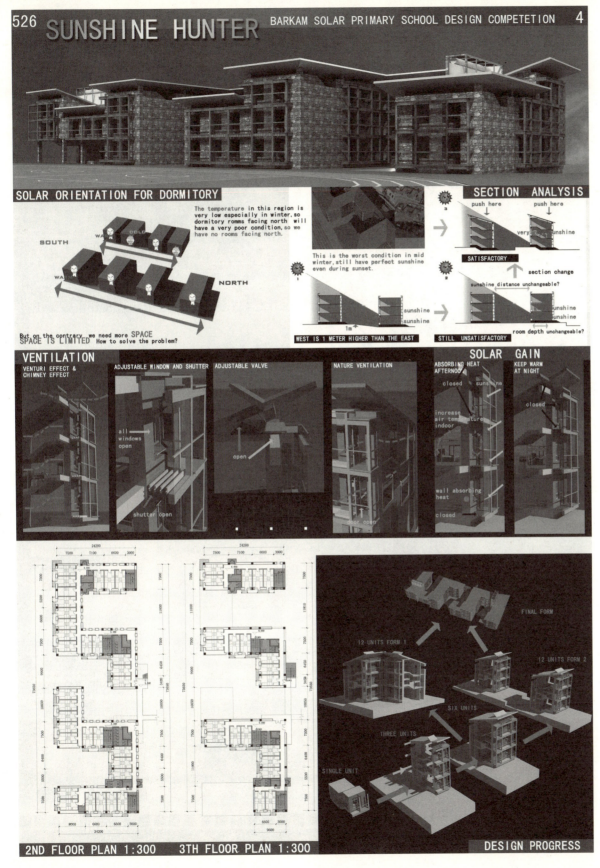

优秀奖
Honorable Mention Prize

项目名称：阳光小学设计
Sunshine Primary School Design

作　　者：郑文崇、谭志臣、林世华、李俊

参赛单位：重庆大学建筑城规学院

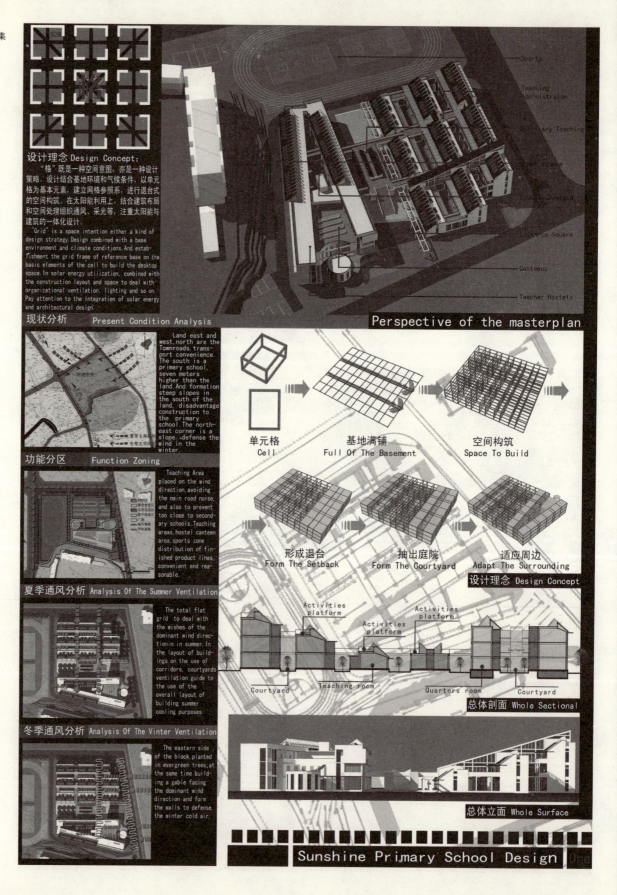

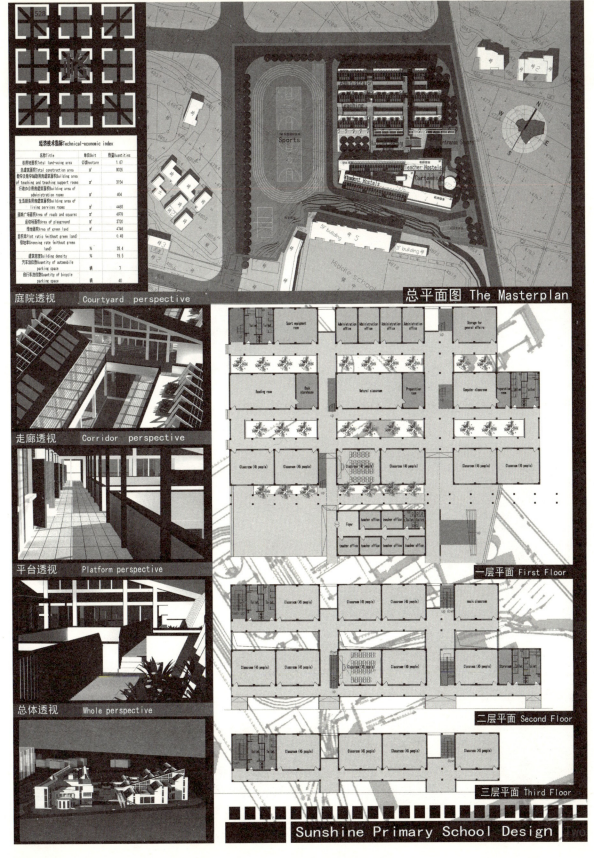

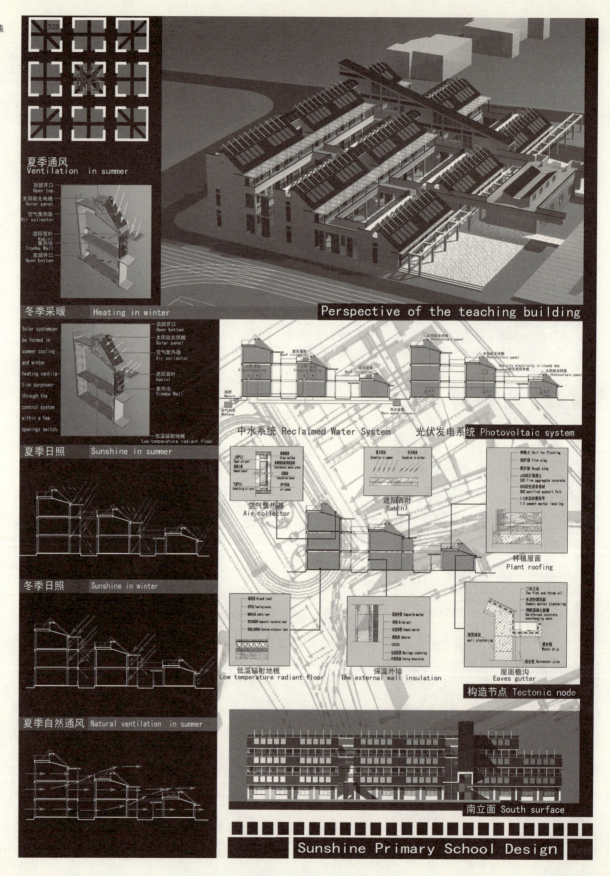

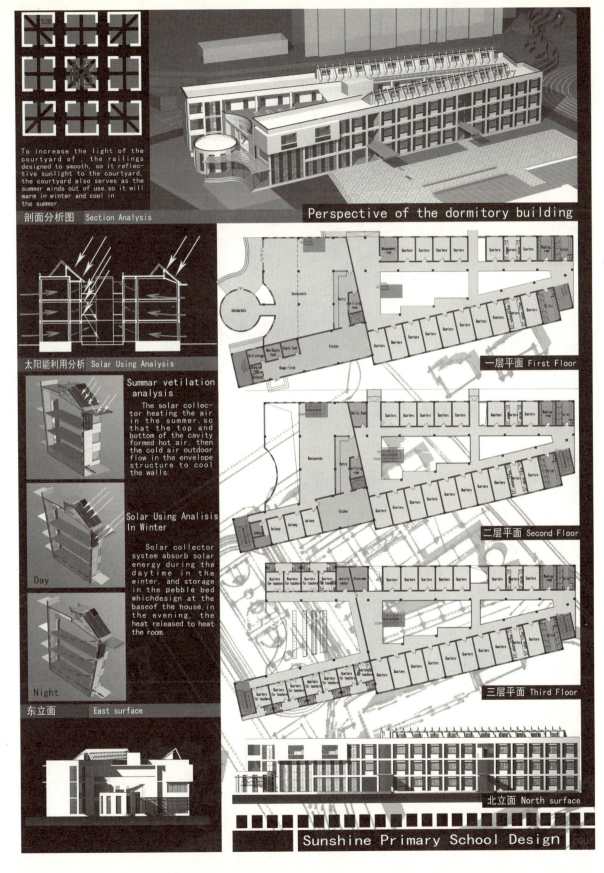

2009台达杯国际太阳能建筑设计竞赛获奖作品集

优秀奖
Honorable Mention Prize

项目名称：534号作品
　　　　　No.534
作　者：祝卿、聂子昊、刘旭明、黄晶、刘晖
参赛单位：华中科技大学建筑与城市规划学院

INTERNATIONAL SOLAR BUILDING DESIGN COMPETITION

REGISTERED CODE: 534　PRE-ANALYSIS

LOCAL CLIMATE ANALYSIS

SOLAR RADIATION (kwh/m²/d) — 4.94, 3.11

Barkam is rich in solar radiation all the year round and belongs to 2nd zone of solar resource.

TEMPERATURE (°C) — 16.2, 15, -0.5

The temperature keeps under the level of heat comfort 15 degrees.

Therefore, the main contradiction of maerkang is to prevent heat losing and to gain heat as much as possible.

RAINFALL (mm) — 160, 0

The period between may to oct. is full of rainfall, increasing the air humidity and raising potential heat. however, the rainfall is poor in other seasons, especially serious in jan.. it leads air humidity raising and increasing the uncomfortable feeling of cold and dry.

EVAPORATION (mm) — 150, 25

This phenomenon is especially serious in the early spring.

LOCAL LOW-TECH COUNTERMEATURES

01 Terrace + out-gallery in south: increasing southern heated area as far as possible, use local solar resource fully.

02

03 Huddling: increasing heat-gained area of inner-surface, decreasing heat-losed area of outer-surface.

04 Tower: as a waring-military construction ago, it degenerates with a culture symble in local today.

REGISTERED CODE: 534   CONCEPT

2009台达杯国际太阳能建筑设计竞赛获奖作品集

从学校功能出发
学校各功能间使用时段分析 ANALYSIS OF EACH FOUNCACITIONAL PERIOD

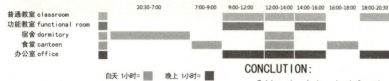

CONCLUTION:
Taking the whole school function in consideration according to the work & rest schedule of the live-in elementary school, the ordinary classrooms and offices are mainly used in daytime while in night. The dormitories are hardly used in the day but fully used at night. Therefore, the thermal insulation rooms such as classrooms, offices & dormitories

THE PHYSQUE FOLLOWS WITH THE SHADOW......
TO GAIN THE MAX. SOLAR REDIATION

1. 根据场地南面建筑一天10点-16点主要太阳辐射热时段所产生的阴影,在其边界外设计宿舍空间
1. Put the dormitory outside the southern shadow boundary of the site made by the sun-heating during 10:00-16:00

2. 再根据南面建筑对二层及其以上的阴影边界,叠加上2、3层宿舍空间
2. Then superposition 2nd 3rd floors according to the shadows affection supplied by southern buildings

3. 在其他可利用空间排布其他功能
3. Arrange other functions in the left usable space

6. 最后在不违反采光前提下排布其他功能空间,完成建筑平面
6. Finally, complete the architectural plane under the condition of following the daylingting requirement

5. 再根据对二层的阴影影响范围,确定二层与二层以上教室与办公功能空间的位置
5. Then arrange the 2nd & higher floors' positions of classrooms & offices considering the shadow affected areas to 2nd floor

4. 根据已经产生的建筑形体,产生的对一层阴影的影响范围,我们确定建筑北面的教室空间
4. The northern classrooms position will be desided according to the shadow affections to 1st floor which are caused by the former constructed buildings

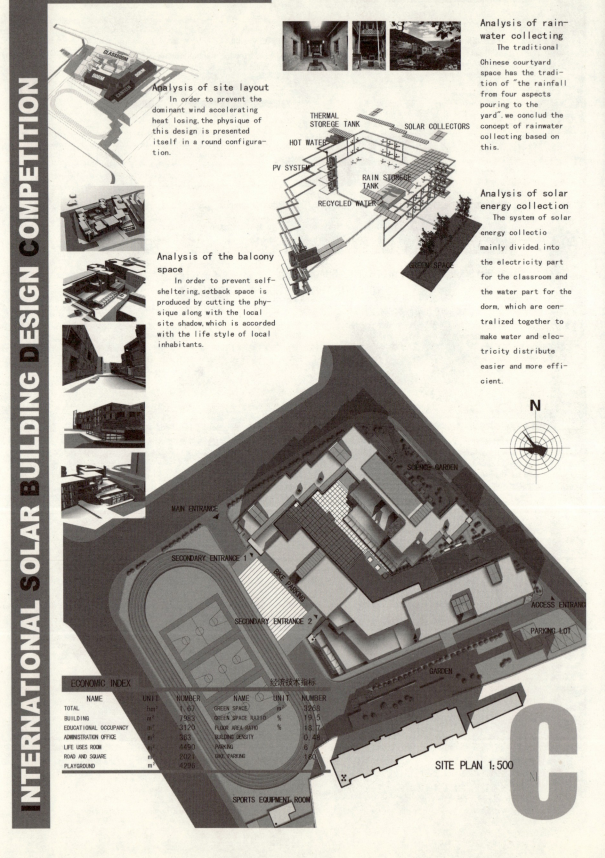

SITE ANALYSIS

REGISTERED CODE: 534

INTERNATIONAL SOLAR BUILDING DESIGN COMPETITION

Analysis of site layout
In order to prevent the dominant wind accelerating heat losing, the physique of this design is presented itself in a round configuration.

Analysis of the balcony space
In order to prevent self-sheltering, setback space is produced by cutting the physique along with the local site shadow, which is accorded with the life style of local inhabitants.

Analysis of rain-water collecting
The traditional Chinese courtyard space has the tradition of "the rainfall from four aspects pouring to the yard", we conclud the concept of rainwater collecting based on this.

Analysis of solar energy collection
The system of solar energy collectio mainly divided into the electricity part for the classroom and the water part for the dorm, which are centralized together to make water and electricity distribute easier and more efficient.

ECONOMIC INDEX / 经济技术指标

NAME	UNIT	NUMBER	NAME	UNIT	NUMBER
TOTAL	hm²	1.67	GREEN SPACE	m²	3268
BUILDING	m²	7983	GREEN SPACE RATIO	%	19.5
EDUCATIONAL OCCUPANCY	m²	3120	FLOOR AREA RATIO	%	18.7
ADMINISTRATION OFFICE	m²	363	BUILDING DENSITY		0.48
LIFE USES ROOM	m²	4490	PARKING		6
ROAD AND SQUARE	m²	2021	BIKE PARKING		180
PLAYGROUND	m²	4296			

SITE PLAN 1:500

REGISTERED CODE: 534   TECHNICAL METHODS

INTERNATIONAL SOLAR BUILDING DESIGN COMPETITION

WINTER DAY
In the day, classroom corridor is fully open to receive solar radiation, gaining heat through Tromble construction in the afternoon meanwhile.

WINTER NIGHT
In the night, the corridor space is sealed by using activated woodplates, preventing heat losing.

SUMMER
In summer, the local solar radiation is strong. These plates can transform into vertical shading, facilitating classroom ventilating.

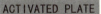

ACTIVATED PLATE

HEAT TOWER (semi-ground)

WINTER DAY
← radiation   Greenhouse-effect
⇐ sunlight   keep energy

WINTER NIGHT
← radiation   hot wall radiat the room.

CHOOSE MATERIAL

SUMMER. AUTUMN
← wind   because of hot pressure, wind appeared.

- Stone: it's effectively for accumulation of heat and it's accessible in local.
- Glass: normal glass is cheap, and made Greenhouse-effect.
- Curtain: it's backside is reflective to reflect the radiation at night, and it's ease of operation.
- Earth contact: earth is effectively for accumulation of heat, and earth face is more cordial for children.

INNER WALL

DRAPE (rolled)
HEAT
LIGHT

In morning, fully accepting solar radiation, the part of accumulating heat begin to work.

DRAPE (rolled)
LIGHT
HEAT

In afternoon, the heat collector begin to release heat gradually, preventing the temperature in classrooms decreasing fiercely.

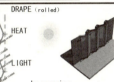

DRAPE (unrolled)
TROMBLE WALL
HEAT

During the night, the tromble construction together with the sealed corridor provide continuous heat to classroom.

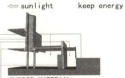

HEAT STORIGE

Day/open

Night/close

The tower In classroom together with the sealed corridor provide continuous heat to classrooms.

HEAT TOWER (classroom)

Rolled up in the day, the windows of dorm is fully opened to solar radiation.

Rolled down in the night, the curtain blocks the heat released from glass.

OPEN

CLOSE

REGISTERED CODE: 534   PLAN ELEVATION SECTION

1. DUTY ROOM
2. HEALTH ROOM
3. TEACHER OFFICE
4. CLASSROOM
5. SCIENCE ROOM
6. PREPARING ROOM
7. INSTRUMENT ROOM
8. MUSIC CLASSROOM
9. COTROLLING ROOM
10. COMPUTER CLASSROOM
11. LABOR CLASSROOM
12. MULTIFUNCIONAL ROOM
13. ART CLASSROOM
14. REALIA ROOM
15. MALE TEACHER BEDROOM
16. BATHROOM
17. MALE STUDENT BEDROOM
18. EQUIPMENT ROOM
19. BOILLING ROOM
20. CANTEEN
21. BATH HOUSE
22. RECEPTION ROOM
23. POWER DISTRIBUTION
24. KIECHEN
25. RESTING ROOM
26. PREPARATION ROOM
27. W.C
28. FEMAIL TEACHER BEDROOM
29. FEMAIL STUDENT BEDROOM
30. SPORTS EQUIPMENT ROOM

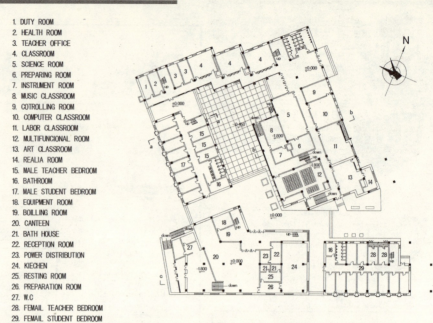

1st FLOOR PLAN 1:300

ELEVATION WEST 1:300

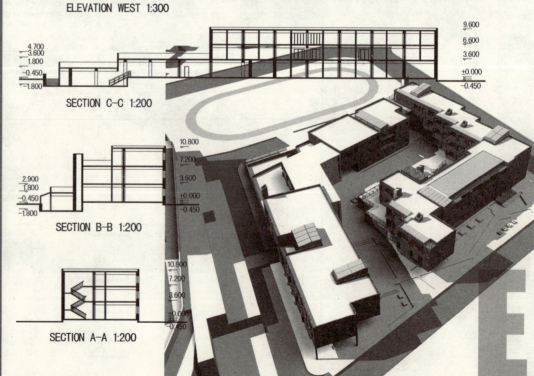

SECTION C-C 1:200

SECTION B-B 1:200

SECTION A-A 1:200

INTERNATIONAL SOLAR BUILDING DESIGN COMPETITION

REGISTERED CODE: 534 PLAN ELEVATION

1. READING ROOM FOR STUDENTS
2. TEACHER OFFICE
3. CLASSROOM
4. ADMINISTRATION OFFICE
5. STORAGE FOR GENERAL AFFAIRS
6. GAZEBO
7. BOOK STOREHOUSE
8. BATHEROOM
9. FEMAILE TEACHER BEDROOM
10. FEMAILE STUDENT BEDROOM
11. SPACE FOR EVACUATE(FEMALE DORM)
12. SPACE FOR EVACUATE(MAILE DORM)
13. MAILE TEACHER BEDROOM
14. MAILE STUDENT BEDROOM
15. STORIGE ROOM

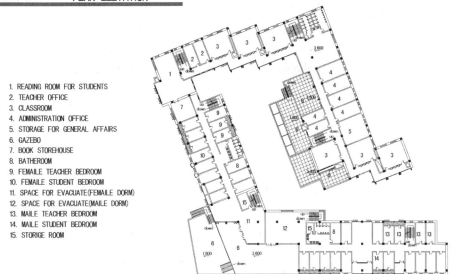

2nd FLOOR PLAN 1:300

ELEVATION EAST 1:300

1. ACTIVE ROOM
2. TEACHER OFFICE
3. CLASSROOM
4. FEMAILE TEACHER BEDROOM
5. BATHEROOM
6. FEMAILE STUDENT BEDROOM
7. SPACE FOR EVACUATE(FEMALE DORM)
8. SPACE FOR EVACUATE(MAILE DORM)
9. STORIGE ROOM
10. MAILE TEACHER BEDROOM
11. MAILE STUDENT BEDROOM

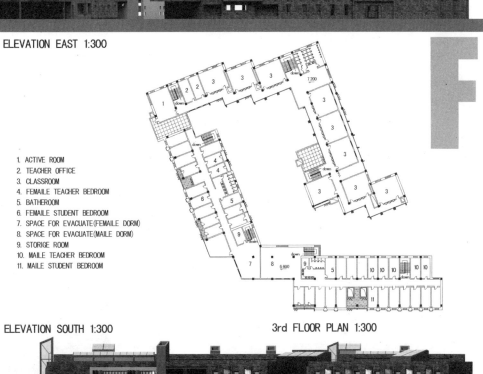

ELEVATION SOUTH 1:300   3rd FLOOR PLAN 1:300

优秀奖
Honorable Mention Prize

项目名称：阳光下的格桑花
　　　　　Sunshine Gesang Flower
作　　者：苏原、邹杰、任娟、李瀛
参赛单位：西北工业大学建筑系

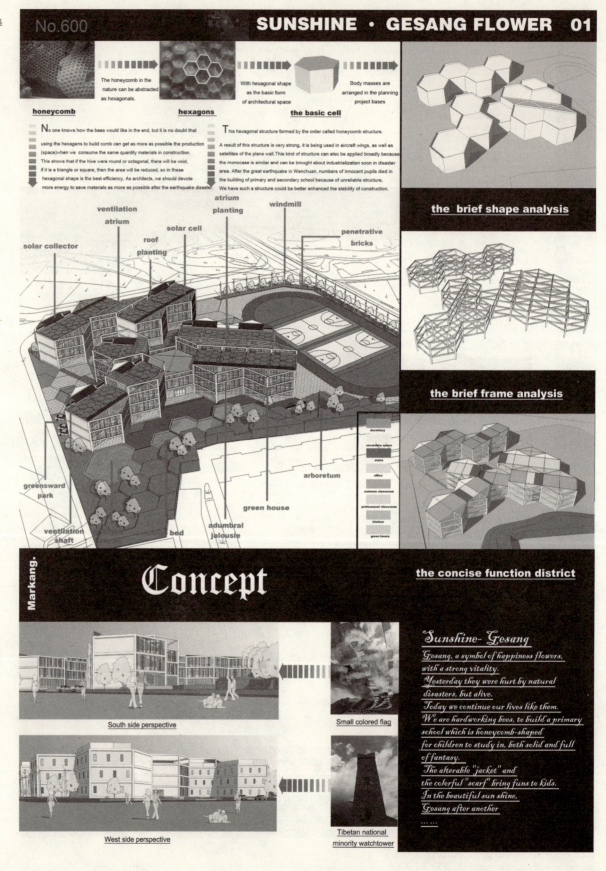

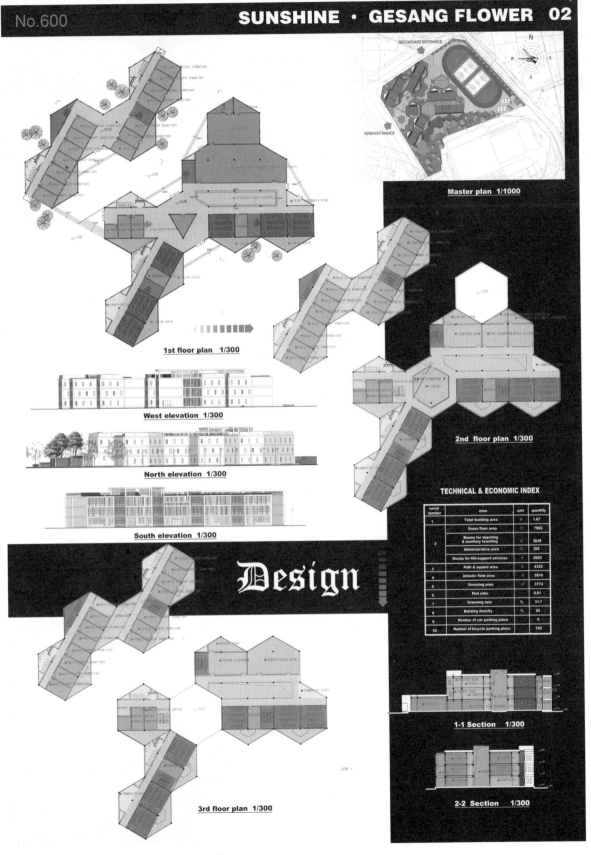

SUNSHINE · GESANG FLOWER 03

No.600

WATER PURIFICATION & RECYCLING SYSTEM

Ecological garden

The ecology garden may maintain the water and soil, adjust humidity of micro-climate. More importantly, it can also make the children to love and protect nature, at last come into being scientific development concept.

The DIY rainwater collecting system

As the image above shows, the concept is very simple. A water butt (rain barrel) is filled directly or via a (or rainwater diverter)

Schematic drawing for MICRO water-cycle

by the rainwater which lands on the roof, runs along the guttering, andflows down a downpipe. This container is then connected to the toilet with a pipe. The whole set up should cost no more than £100 and a couple of hours of your time.

The rainwater collecting & purification constitution

VENTALATION IN SUMMER

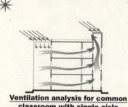

Ventilation analysis for common classroom with single aisle

In the daytime, turning on all of the windows, the fresh air blows from the northwest direction forms the through-draught. When temperature increment, the

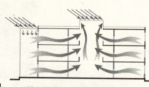

Ventilation analysis for common classroom with atrium

sunlight shines on the jellaba causes the partial air heating to form the thermal pressure in the solar chimney to ventilate. At night, the northwestern weather windows have been shut up and that the thermal pressure play an importan role in ventilation.

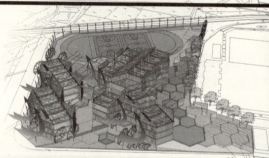

VENTALATION IN WINTER

Ventilation analysis for common classroom with single aisle

The cold and dry air is heat up and damped by green house then draughted into the common classroom or teachers offices.When the sunshine heats the

Ventilation analysis for common classroom with atrium

dark jellaba in order to form the thermal pressure.In the atrium,several kinds of plants can also make the air fresh and clean.

Eco-strategy

DAYLIGHTING ANALYSIS

The morning daylighting analysis

When the solar altitude is low in the morning, then the angle of jalousie and t horizotal line is small, therefore the sunshine may penetrate the persiennes make the inner space of the classroom clear, and cause the horizontal plane of the classroom to recieve the degree of illumination quite even.

The midday daylighting analysis

When the solar altitude is high in the midday, we adjust the angle of jalousie and t horizotal line to make it raised, therefore the sunshine may not penetrate the persiennes. In this way, the space near the windows cannot be glare to kids' reading and writing, cause the horizontal plane of the classroom to clear other than glazing.

The afternoon daylighting analysis

When the solar altitude is lower and lower in the afternoon, then the angle of jalousie and t horizotal line become small again, therefore the sunshine may penetrate the persiennes make the inner space of the classroom clear, and cause the horizontal plane of the classroom bright.

WIND ENERGY SYSTEM

The main benefit to switching to wind power is the improvement to the environment. Wind power can displace power from fossil-fueled power plants and help to improve on greenhouse gas emissions that are hurting our environment.

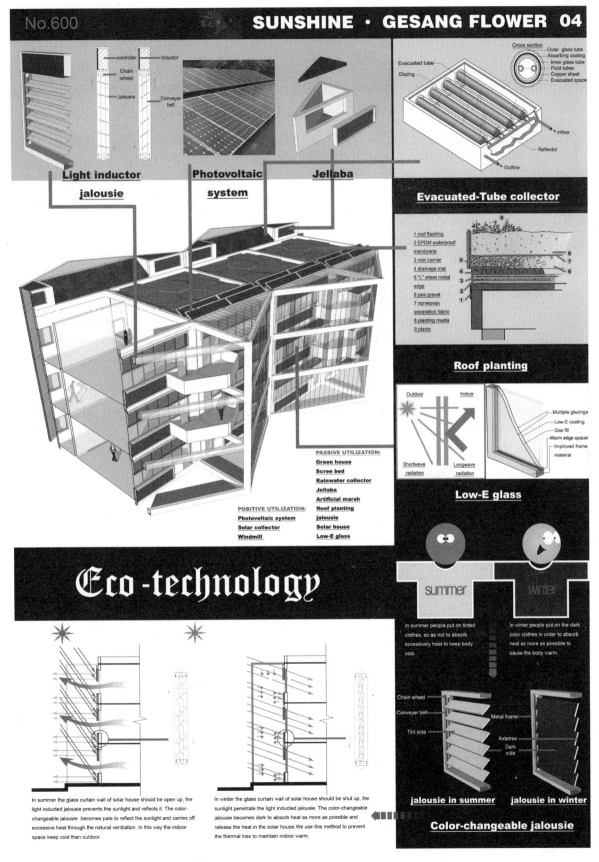

优秀奖
Honorable Mention Prize

项目名称：太阳-斜面
　　　　　Solar-Slope
作　　者：夏君天、张愉
参赛单位：清华大学建筑学院

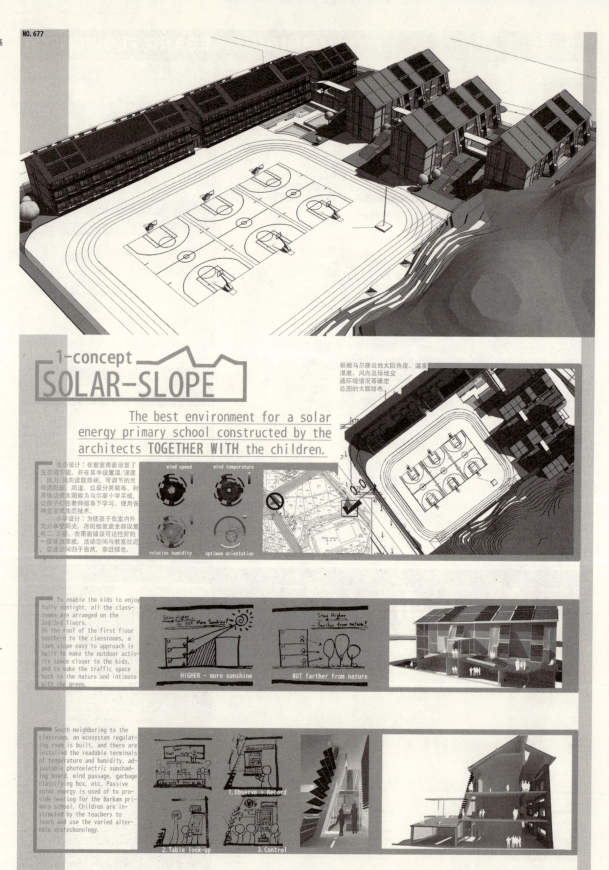

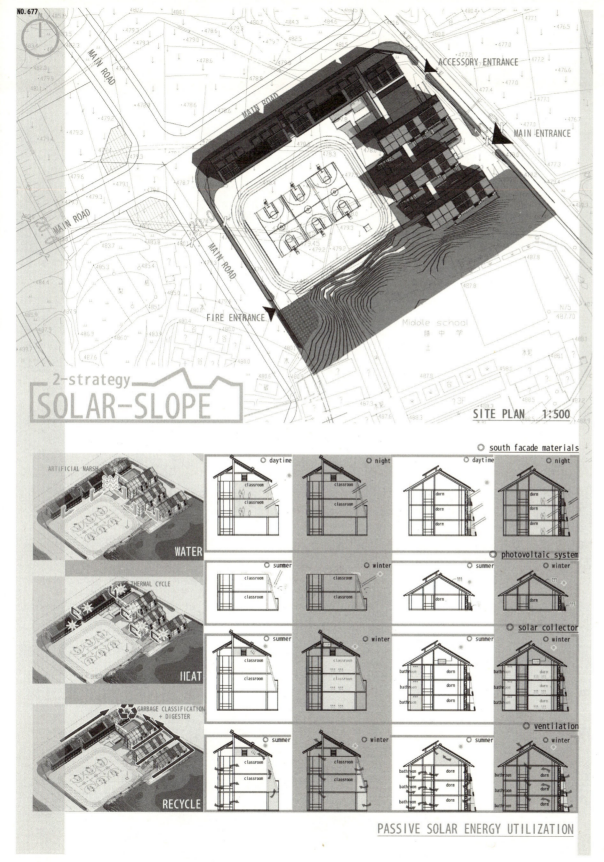

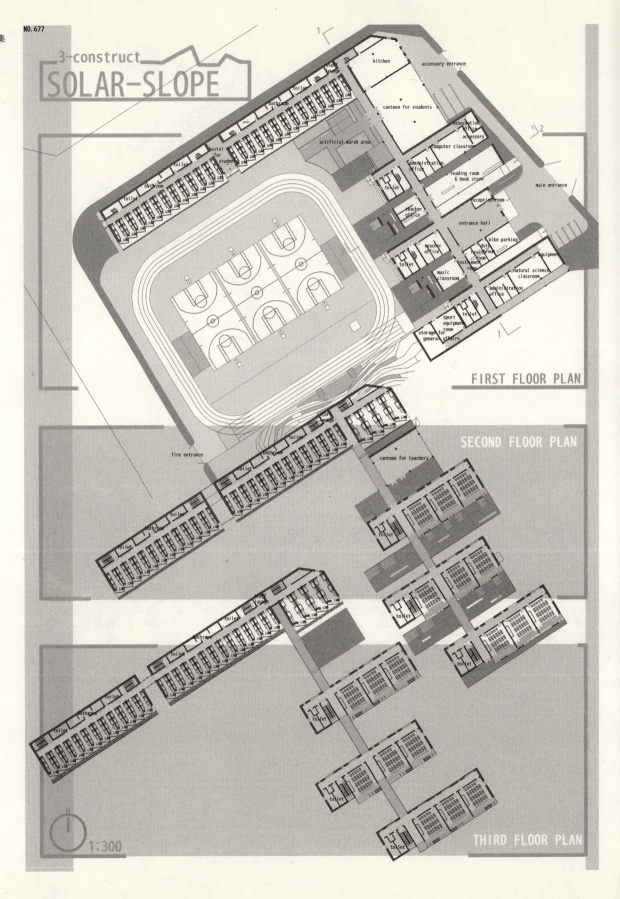

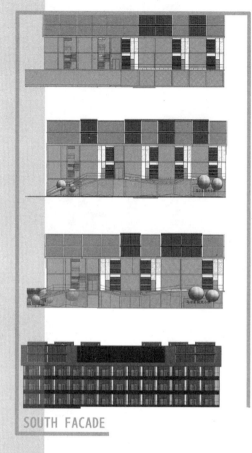

SOUTH FACADE

TEACHING AREA

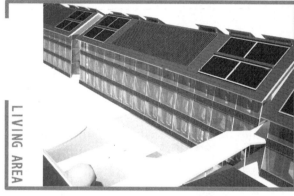
LIVING AREA

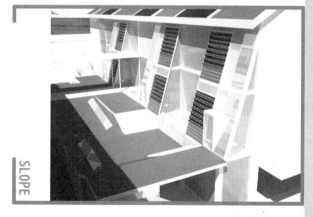
SLOPE

4-appearance
SOLAR-SLOPE

WINTER-WIND SPEED DROPPED   SUMMER-WELL VENTILATED

ECONOMIC ANALYSIS

ITEM	UNIT	AMOUNT
TOTAL AREA	hm²	1.67
BUILDING AREA	m²	7950
TEACHING AREA	m²	3200
OFFICE AREA	m²	300
LIVING AREA	m²	4450
ROADWAY&SQUARES	m²	1500
SCHOOLYARD	m²	3100
GREENBELT	m²	4300
FLOOR AREA RATIO		0.47
GREENING RATE	%	25
BUILDING DENSITY	%	21
AUTO BERTH		6
BICYCLE BERTH		80

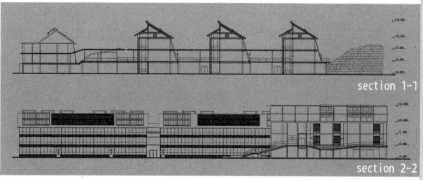

section 1-1

section 2-2

优秀奖
Honorable Mention Prize

项目名称：分享阳光
　　　　　Share Sunshine

作　　者：万珊、宋迎

参赛单位：华中科技大学

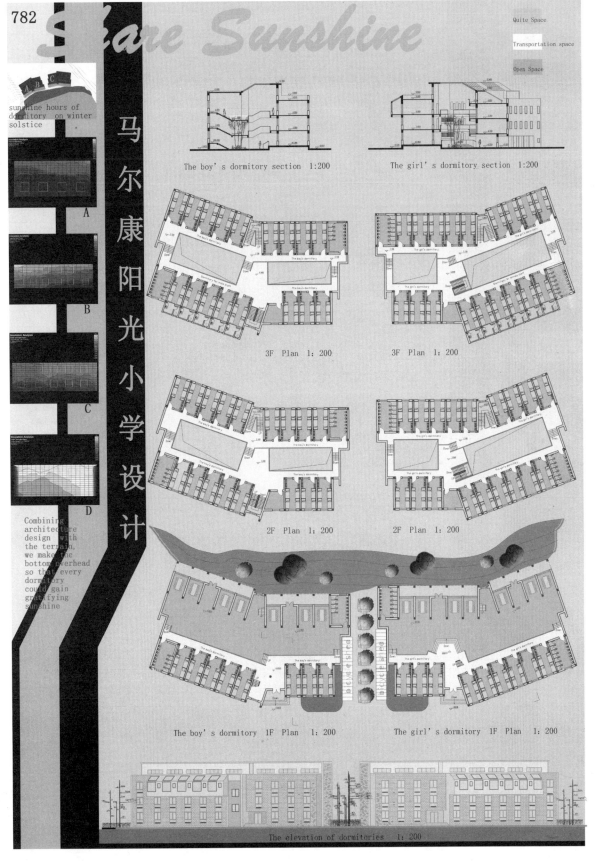

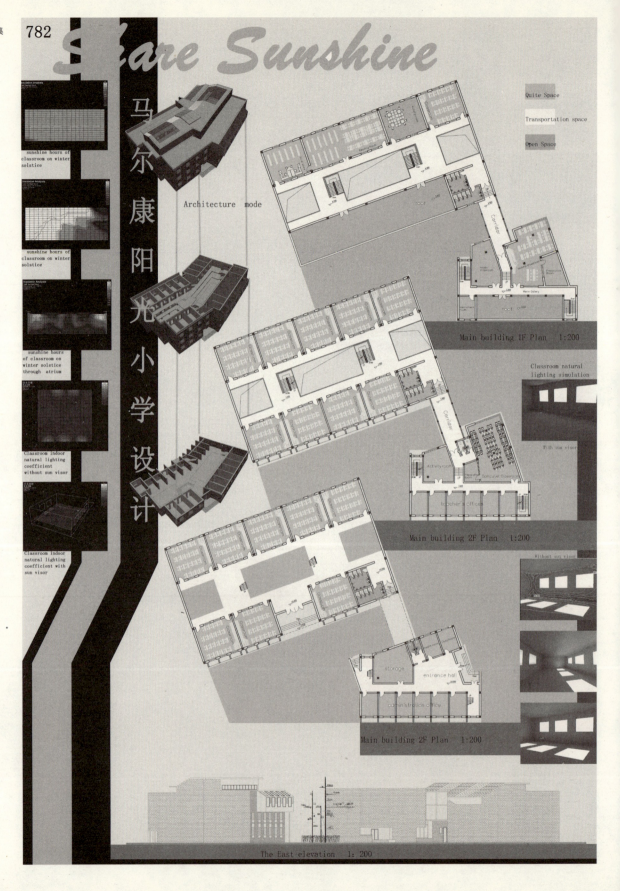

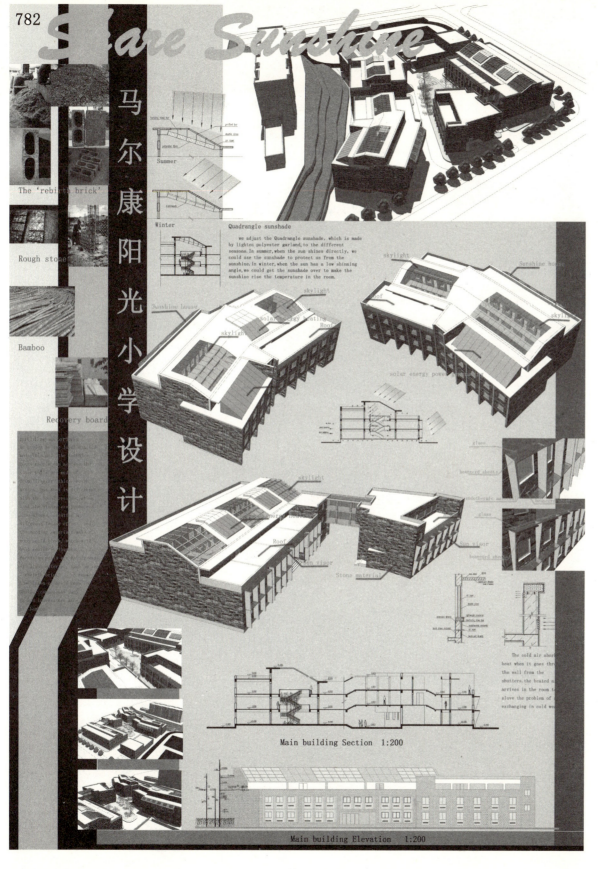

优秀奖
Honorable Mention Prize

项目名称：阳光·乐园
　　　　　Sunny Garden School
作　　者：任乃鑫、王磊、谢欢欢、毕岩、
　　　　　张轲、程广红
参赛单位：沈阳建筑大学建筑与规划学院

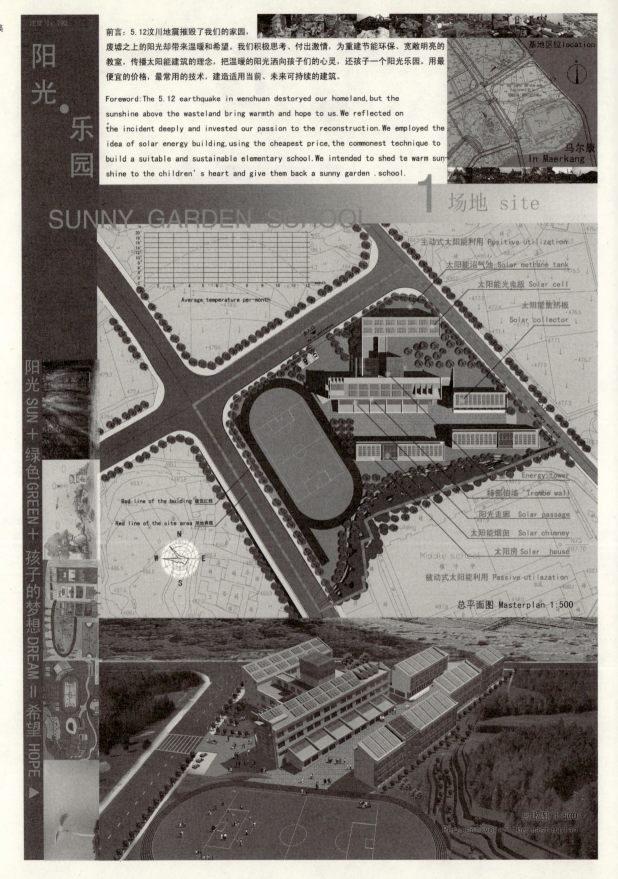

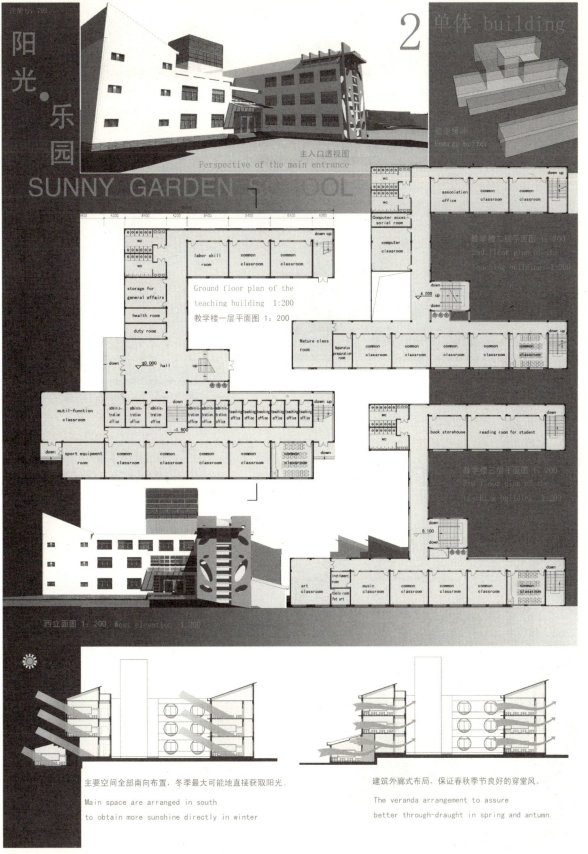

阳光·乐园 SUNNY GARDEN SCHOOL

2009台达杯国际太阳能建筑设计竞赛获奖作品集

2 单体 building

主入口透视图 Perspective of the main entrance

能量缓冲 Energy buffer

教学楼二层平面图 1:200
2nd floor plan of the teaching building 1:200

Ground floor plan of the teaching building 1:200
教学楼一层平面图 1:200

教学楼三层平面图 1:200
3rd floor plan of the teaching building 1:200

西立面图 1:200 West elevation 1:200

主要空间全部南向布置，冬季最大可能地直接获取阳光。
Main space are arranged in south to obtain more sunshine directly in winter

建筑外廊式布局，保证春秋季节良好的穿堂风。
The veranda arrangement to assure better through-draught in spring and antumn.

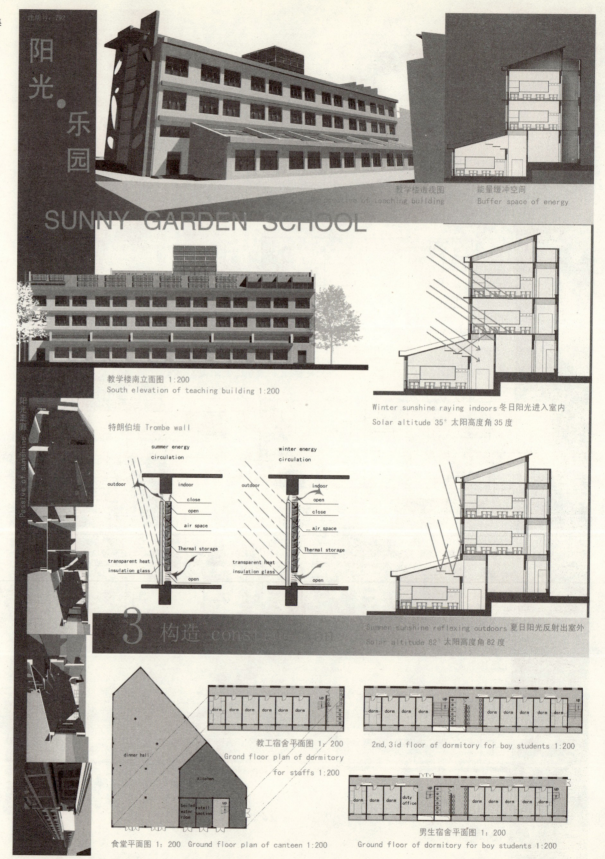

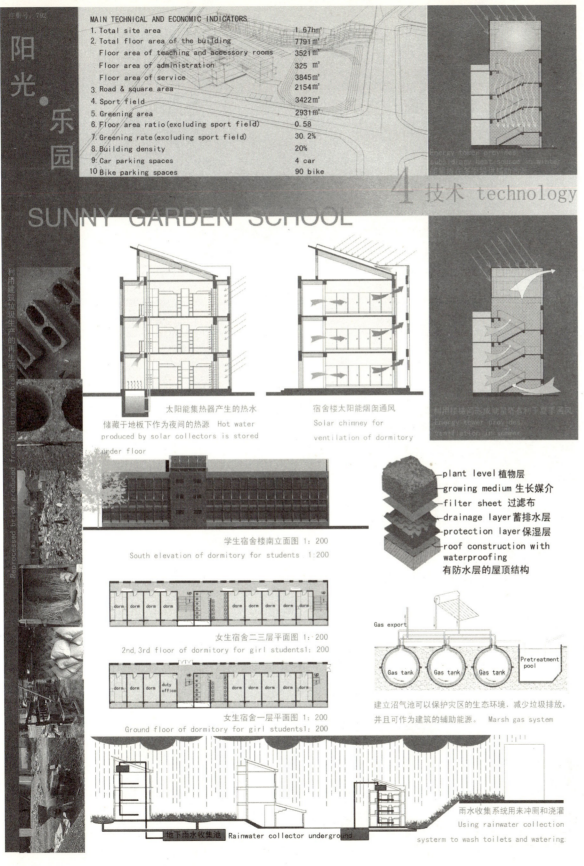

优秀奖
Honorable Mention Prize

项目名称：阳光街小学
　　　　　The Solar Road

作　　者：Tjerk Reijenga、蔡晓琦、张嫣、
　　　　　宋慧、郑欣、贺玮玮、杨丽、陈鸣

参赛单位：KOW International Dutch Design Consultants 凯维建筑设计咨询(上海)有限公司

THE SOLAR ROAD

阳光街小学 794

Ma'erkang Area, Sichuan Province
31.5 °N, 102.2 °E altitude 2666 m.

Mountain Climate:
Cold winters/
Mild summers/
Plentiful sunshine

设计灵感来源于马尔康传统的建筑形式：天然毛石材料、泥砖，向阳建筑组团内的木构庭院空间。设计中除了最大化利用太阳能之外，我们还尽可能多地运用当地材料和建造技术。考虑到当地的实际情况，建筑设计在经济上必须是可行的：非高技派的太阳能设计。
基地对于所要求的面积显得很小，并且受南边山坡上建筑阴影的影响。为了在早上能够得到最多的太阳能，建筑需位于基地的东北边。
有两个主要的建筑体块：宿舍楼和教学楼。供低年级学生用的教室则分开设置。
为了在早上得到更多的太阳能教学楼被相应的抬高。为获取更多的日照，宿舍楼则置于餐厅上方。玻璃屋顶覆盖了内部中庭并形成了太阳能中庭，这一中庭的热性能有助于调节室内气候环境。

ARCHITECTURAL CONCEPT

The architecture is inspired by the traditional architecture of Ma'erkang: robust natural stone, adobe, softer wood-covered spaces in the courtyards with individual buildings in groups that face the sun. Besides maximizing the use of solar features, our design sets out to use mostly local materials and building techniques. In this rural setting, the design should be economically feasible: solar design without high-tech.

The site is very small for the requested functions and shaded by the tall building and the hill on the south. The buildings are located in the north-east edge of the site for maximum solar gain in the mornings.

There are two main building volumes: the dormitories and the classroom buildings. The classrooms for the youngest kids are separated from the others.

The classroom building is lifted to receive more solar gain in the early morning. The dormitories are stacked on top of the canteen to gain more solar energy.

The internal street is covered by a glass roof and becomes the SOLAR ROAD. The heat-gaining capacity of this street helps regulate the indoor climate.

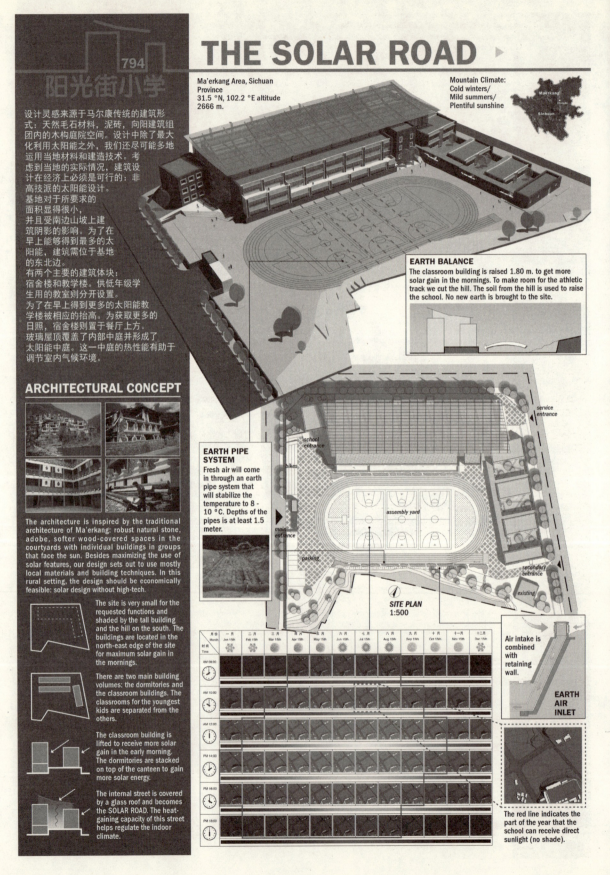

EARTH BALANCE
The classroom building is raised 1.80 m. to get more solar gain in the mornings. To make room for the athletic track we cut the hill. The soil from the hill is used to raise the school. No new earth is brought to the site.

EARTH PIPE SYSTEM
Fresh air will come in through an earth pipe system that will stabilize the temperature to 8 - 10 °C. Depths of the pipes is at least 1.5 meter.

SITE PLAN 1:500

Air intake is combined with retaining wall.

EARTH AIR INLET

The red line indicates the part of the year that the school can receive direct sunlight (no shade).

THE SOLAR ROAD

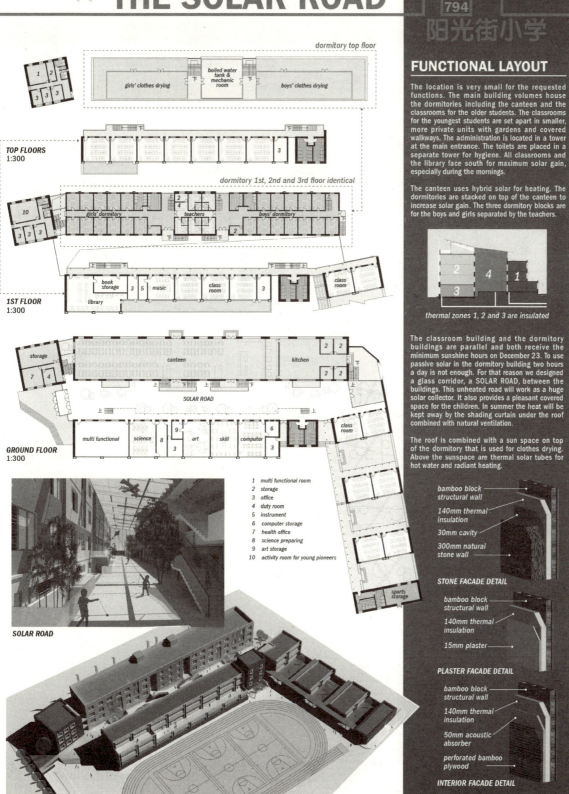

FUNCTIONAL LAYOUT

The location is very small for the requested functions. The main building volumes house the dormitories including the canteen and the classrooms for the older students. The classrooms for the youngest students are set apart in smaller, more private units with gardens and covered walkways. The administration is located in a tower at the main entrance. The toilets are placed in a separate tower for hygiene. All classrooms and the library face south for maximum solar gain, especially during the mornings.

The canteen uses hybrid solar for heating. The dormitories are stacked on top of the canteen to increase solar gain. The three dormitory blocks are for the boys and girls separated by the teachers.

thermal zones 1, 2 and 3 are insulated

The classroom building and the dormitory buildings are parallel and both receive the minimum sunshine hours on December 23. To use passive solar in the dormitory building two hours a day is not enough. For that reason we designed a glass corridor, a SOLAR ROAD, between the buildings. This unheated road will work as a huge solar collector. It also provides a pleasant covered space for the children. In summer the heat will be kept away by the shading curtain under the roof combined with natural ventilation.

The roof is combined with a sun space on top of the dormitory that is used for clothes drying. Above the sunspace are thermal solar tubes for hot water and radiant heating.

- bamboo block structural wall
- 140mm thermal insulation
- 30mm cavity
- 300mm natural stone wall

STONE FACADE DETAIL

- bamboo block structural wall
- 140mm thermal insulation
- 15mm plaster

PLASTER FACADE DETAIL

- bamboo block structural wall
- 140mm thermal insulation
- 50mm acoustic absorber
- perforated bamboo plywood

INTERIOR FACADE DETAIL

THE SOLAR ROAD ▶▶▶

SOLAR DESIGN STRATEGY

Considering the climate conditions, the focus will be on heating demand. Cooling is not necessary if the buildings are designed for the climate. Comfort temperature is 16-18 °C (lower than Western standards).

The SOLAR ROAD is an unheated street between the classroom building and the dormitory building. The (security)glass roof will gain a lot of solar irradiation. On average, the temperature in the Solar Road will be 10 °C higher than the outside temperature.

Solar Road [Winter]
Fresh air will come into the solar road through an earth pipe system at 8 - 10 °C. In the solar road this air can heat up another 10 °C by solar gain. During the daytime in winter the indoor temperature will be rather comfortable.

Solar Road [Summer]
In the summer, the semi-transparent curtain below the glass roof will be closed (25% of daylight will still come through). The air gaps between glass roof and curtain will be opened on the lowest point for fresh air and the hotter air will leave on the highest point int into the solar clothes drying area. This prevents overheating.

Canteen and Kitchen [Winter]
Solar gain will enter at the end of the morning through the window in the south façade. Additional heating is needed with thermal solar collectors on the roof. The low temperature heat will be distributed with a radiant floor heating system. The earth pipe system supplies fresh air at 10 °C. Mechanical ventilation in the kitchen above the stoves will suck air out of the kitchen.

Canteen and Kitchen [Summer]
The earth pipe system provides pre-cooled air at 10 °C. Fresh air for the kitchen will come from the canteen.

Dormitory [Winter]
The south side of the dormitory building gains passive solar energy. The heating demand is mainly in the afternoon and evening. Air ducts lead warm air from the solar road to the north dormitories. Shunt-shaft chimneys will suck the air out of the bath- and bedrooms. These are equipped with wind-catchers, providing ventilation without a mechanical system.

Dormitory [Summer]
The air in the solar road is protected from overheating by the semi-transparent curtain. Pre-cooled fresh air will come into the solar road through the earth pipe system.

Administration [Winter]
Solar gain will heat up the offices on the south facade. Solar collectors provide additional heating, distributed through a radiant floor heating system.

Administration [Summer]
Overheating is prevented with outside shading louvres. Pre-cooled air comes in through the earth pipe system.

solar hot water
bamboo structural beam
aluminum greenhouse construction system
floors:
- laminated bamboo ceiling beams
- bamboo plywood
- adobe floor
- bamboo laminated flooring

interior finish:
- white painted plaster
- bamboo plywood cladding

natural stone thermal insulation (XPS) with plaster 140mm

green roof

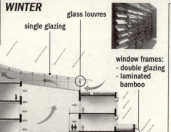

WINTER

single glazing
glass louvres

window frames:
- double glazing
- laminated bamboo

WIND VANE
The wind vane pivots horizontally on a rotating mount. The natural wind provides the ventilation and sucks the air from the dormitories below.

wind
indoor air

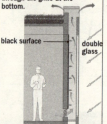

SOLAR CHIMNEY
Solar chimneys work both on sunny days and in overcast conditions. Air heats up in the black vertical glazed shaft escaping through the adjustable grille in the top. As it escapes, the air sucks indoor air from the classrooms through the grille at the bottom.

black surface
double glass

SUMMER

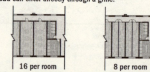

ALTERNATIVE DORMITORY LAYOUT
The floor plans show a traditional dormitory floorplan with a central corridor and eight students to a room – as stated in the competition brief. Alternative lay-outs (below) for the dormitory avoid bedrooms along the north façade. Air ducts become unnecessary and warm air from the solar road can enter directly through a grille.

16 per room | 8 per room

ENTRANCE ELEVATION

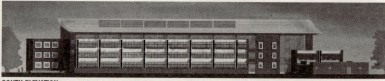

SOUTH ELEVATION

vent
entrance

NORTH ELEVATION

THE SOLAR ROAD

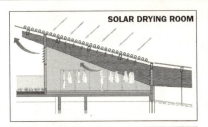

SOLAR DRYING ROOM

BAMBOO BEAM
The main structural beams that carry the roof are made of laminated bamboo reinforced with internal bamboo flanges.

Rainwater flows from the glass roof into a gutter on the roof of the classroom building. It is collected in a reservoir in the top of the toilet block where it can be used for flushing and cleaning.

OPTIONAL PHOTO VOLTAIC LAMINATES
The glass roof above the solar road can be used for semi-transparent photovoltaic laminates in the future. The gained electricity can be supplied to the local electricity grid or directly used.

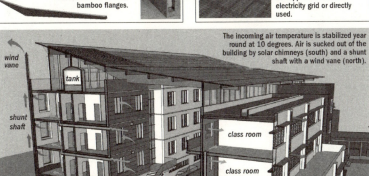

The incoming air temperature is stabilized year round at 10 degrees. Air is sucked out of the building by solar chimneys (south) and a shunt shaft with a wind vane (north).

VENTILATION

CLASSROOM DAYLIGHT DISTRIBUTION
The amount of direct and diffuse daylight in the classroom can be adjusted with external lamellas at the south window. An internal light shelf helps spread the day light more evenly throughout the room. An external overhang blocks out the strong summer sun.

MAIN TECHNICAL AND ECONOMIC INDICATORS

Serial number	Item	Unit	Quantity
1	Total site area	ha	1.67
2	Total floor area of the building	m²	7998
	Floor area of teaching and accessory rooms	m²	2881
	Floor area of administration	m²	418
	Floor area of service	m²	4699
3	Road & square area	m²	5061
4	Sport field	m²	3574
5	Greening area	m²	3934
6	Floor area ratio(excluding sport field)		0.6
7	Greening rate(excluding sport field)	%	30
8	Building density	%	25
9	Car parking spaces	car	6
10	Bike parking spaces	bike	90

Blinds prevent too much daylight in the classrooms. The curtain under the glass roof is closed.
DAYLIGHT SUMMER

Blinds are open and daylight can penetrate deep into the building.
DAYLIGHT WINTER

Hot air is sucked out through the vents under the glass roof. The curtain is closed. Cool air comes in through the earth pipes.
SOLAR SUMMER

The SOLAR ROAD heats up pre-heated air from the earth pipes. The warm air is used to heat up the dormitories.
SOLAR WINTER

CALCULATIONS

In order to analyse and verify the energy and daylighting claims and assumptions in our school design, we have done a series of Leed Energy Simulation analyses on our building design. The aspects we are most interested in are the daylight values in the classrooms and dormitory spaces, the solar gain and temperature in the dormitories – especially in the evenings – and potential ventilation/draft problems in the SOLAR ROAD due to wind.

We have used CFD and other eco-calculation software to check our results. The diagrams and figures below illustrate some of our findings and conclusions for the study on energy and daylighting efficiency of the buildings.

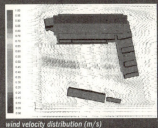
wind velocity distribution (m/s)

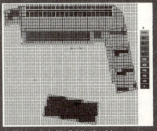
daylight analysis (daylighting levels)

CONCLUSIONS

1 Based on the CFD simulation results, the wind pressure drop on the building surfaces is too low to induce the gust at gate. So door/opening can be located at will or according to other demands. In addition, because of the configuration of green plant belt and the building, the area with high wind velocity is located in the belt zone from the west entrance to the east of the playground.

2 Based on the daylighting analysis result, the daylight facor and level under this configuration of windows is acceptable according to GB/T 50033—2001. The daylight factor and level is much higher than the criterion and the daylight level is more than 300lx in most places.

3 Based on the dynamic energy simulation results, much of the daytime heat in the dormitory is lost at night through the windows. To solve this, dormitory windows should be equipped with shutters to reduce heat loss and contain heat during the night.

优秀奖
Honorable Mention Prize

项目名称：阳光容器
A Container Of Sunshine

作 者：刘铸、王强、张建伟
参赛单位：东南大学建筑学院

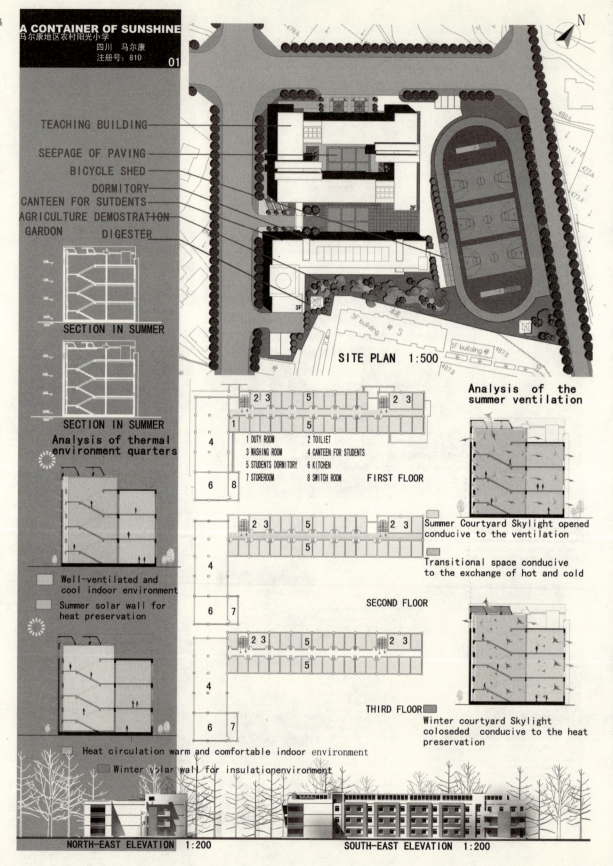

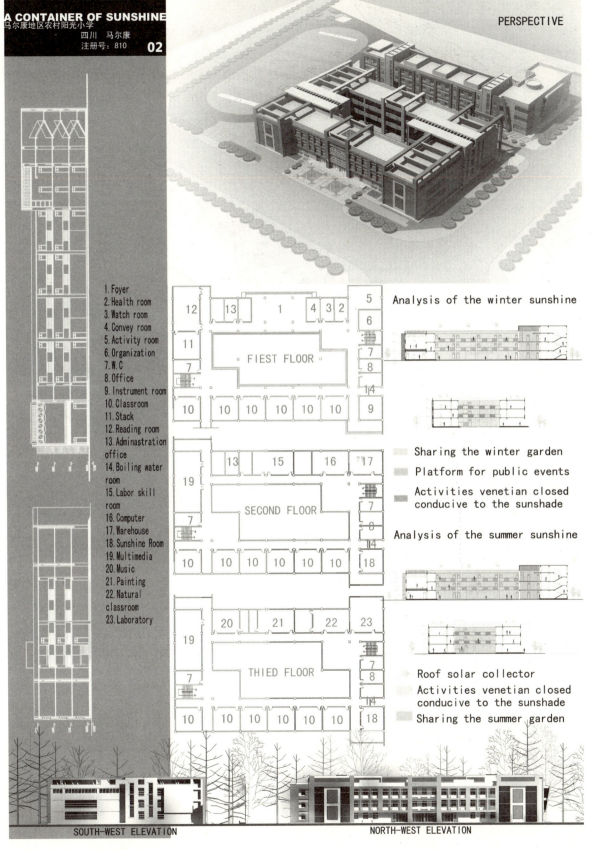

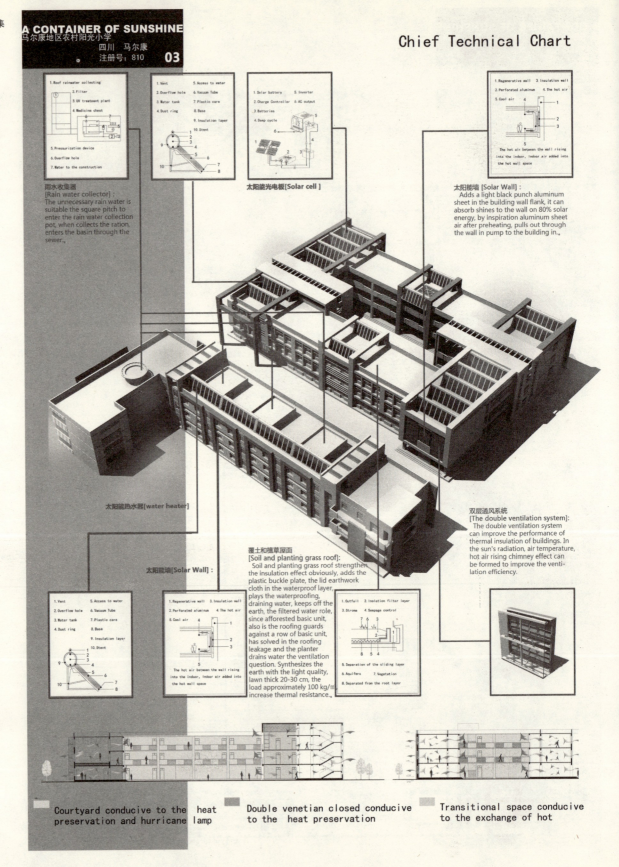

A CONTAINER OF SUNSHINE

马尔康地区农村阳光小学
四川 马尔康
注册号：810

04

设计说明：
设计注重当地生态，以被动式技术为主，注意太阳能利用、雨水收集，优先考虑良好的日照与通风，结合地方特色与生态技术手段，为小学生创造舒适宜人的学习生活环境。

Design Description:
Design focus on the local ecology, to passive technology-based, pay attention to solar energy, rainwater collection, to give priority to good sunshine and ventilation, combined with local characteristics and ecological technology for primary and secondary school students learning to create comfortable and pleasant living environment.

The main economic and technical indicators:

The total land area: 16700m2
Total construction area: 7970m2
Teaching space construction area: 3670m2
Administrative space construction area: 370m2
Living space floor area: 3930m2
Square area of the road: 3402m2
Sports area: 4300m2
Green area: 5998m2

Teaching building perspective

Quarters perspective

Teaching building perspective

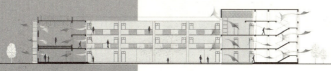

Courtyard conducive to the sunshade

Deep shade and open space conducive to the exchange of hot

2009 台达杯国际太阳能建筑设计竞赛获奖作品集

优秀奖
Honorable Mention Prize

项目名称：感光
Sense The Sunlight
作　者：易勇、刘子锷、区志勇、兰兵、鲍帆
参赛单位：武汉大学城市设计学院建筑系

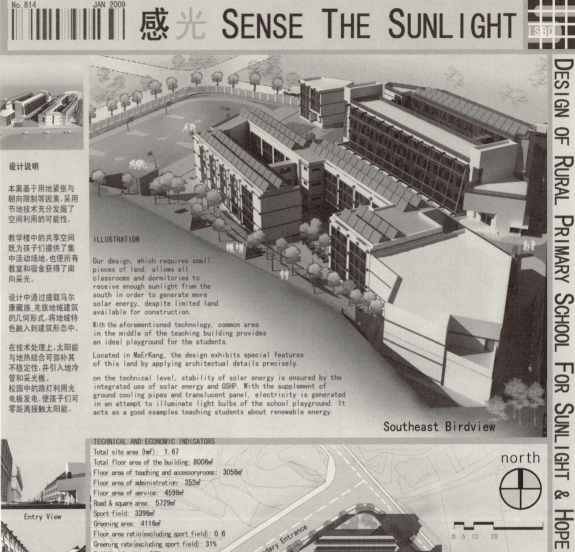

感光 SENSE THE SUNLIGHT

设计说明

本案基于用地紧张与朝向限制等因素，采用节地技术充分发掘了空间利用的可能性。

教学楼中的共享空间既为孩子们提供了集中活动场地，也使所有教室和宿舍获得了南向采光。

设计中通过提取马尔康藏族、羌族地域建筑的几何形式，将地域特色融入到建筑形态中。

在技术处理上，太阳能与地热结合可弥补其不稳定性，并引入地冷管和采光板。

校园中的路灯利用光电板发电，使孩子们可零距离接触太阳能。

ILLUSTRATION

Our design, which requires small pieces of land, allows all classrooms and dormitories to receive enough sunlight from the south in order to generate more solar energy, despite limited land available for construction.

With the aforementioned technology, common area in the middle of the teaching building provides an ideal playground for the students.

Located in MaErKang, the design exhibits special features of this land by applying architectual details precisely.

on the technical level, stability of solar energy is ensured by the integrated use of solar energy and GSHP. With the supplement of ground cooling pipes and translucent panel, electricity is generated in an attempt to illuminate light bulbs of the school playground. It acts as a good examples teaching students about renewable energy.

Southeast Birdview

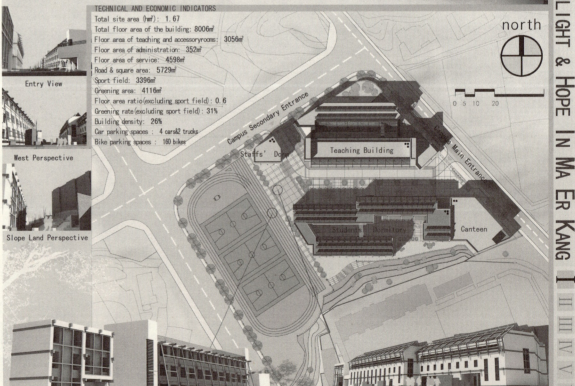

TECHNICAL AND ECONOMIC INDICATORS

Total site area (hm²): 1.67
Total floor area of the building: 8006m²
Floor area of teaching and accessoryrooms: 3056m²
Floor area of administration: 352m²
Floor area of service: 4598m²
Road & square area: 5729m²
Sport field: 3396m²
Greening area: 4116m²
Floor area ratio(excluding sport field): 0.6
Greening rate(excluding sport field): 31%
Building density: 26%
Car parking spaces: 4 cars&2 trucks
Bike parking spaces: 160 bikes

Entry View
West Perspective
Slope Land Perspective

Site Plan 1:500

DESIGN OF RURAL PRIMARY SCHOOL FOR SUNLIGHT & HOPE IN MA ER KANG

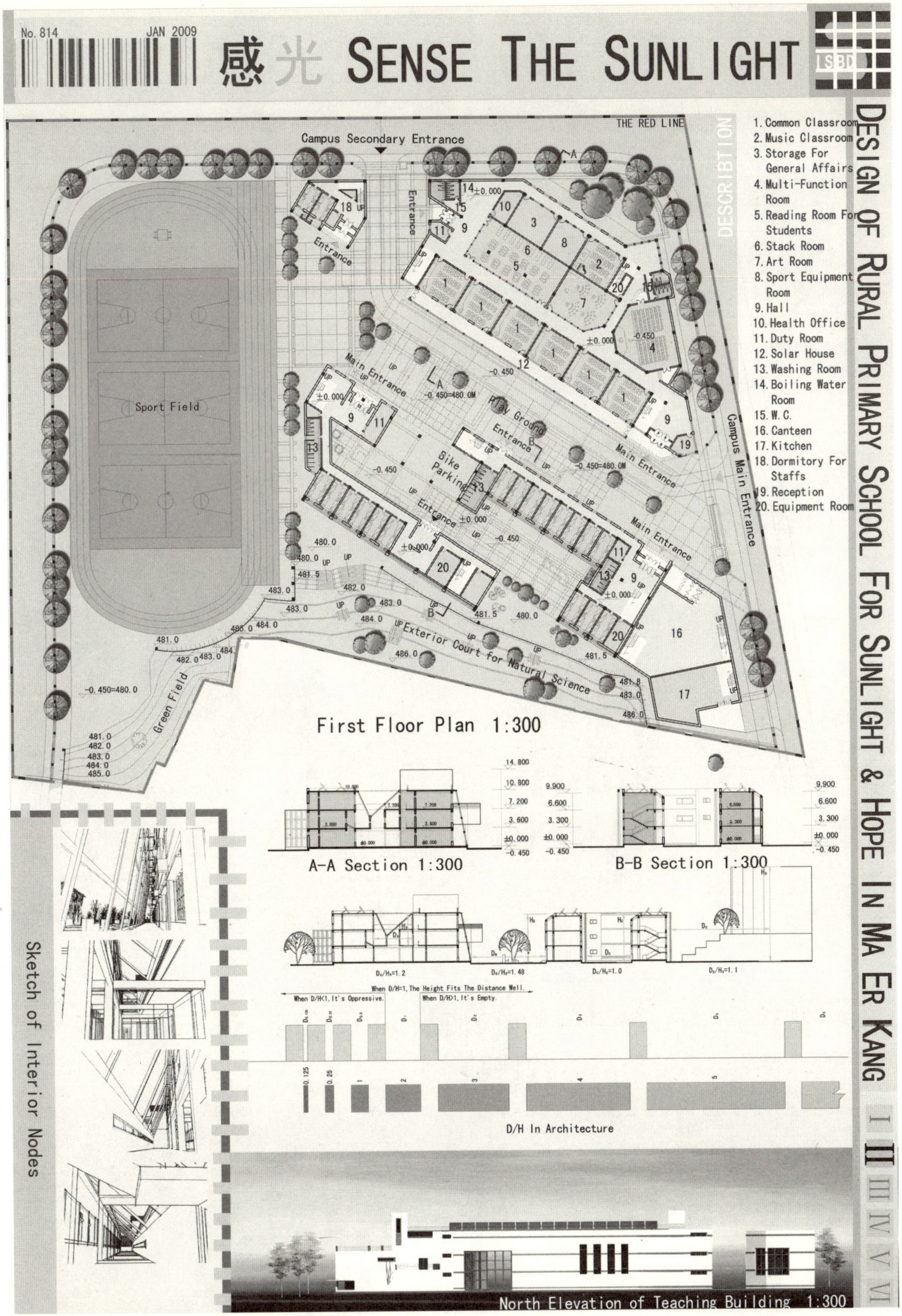

No. 814
JAN 2009
感光 SENSE THE SUNLIGHT
ISBD
DESIGN OF RURAL PRIMARY SCHOOL FOR SUNLIGHT & HOPE IN MA ER KANG
DESCRIBTION
1. Common Classroom
2. Music Classroom
3. Storage For General Affairs
4. Multi-Function Room
5. Reading Room For Students
6. Stack Room
7. Art Room
8. Sport Equipment Room
9. Hall
10. Health Office
11. Duty Room
12. Solar House
13. Washing Room
14. Boiling Water Room
15. W.C.
16. Canteen
17. Kitchen
18. Dormitory For Staffs
19. Reception
20. Equipment Room
THE RED LINE
Campus Secondary Entrance
Campus Main Entrance
Main Entrance
Entrance
Play Ground
Bike Parking
Sport Field
Green Field
Exterior Court for Natural Science
First Floor Plan 1:300
A-A Section 1:300
B-B Section 1:300
When D/H=1, The Height Fits The Distance Well.
When D/H<1, It's Oppressive.
When D/H>1, It's Empty.
D/H In Architecture
Sketch of Interior Nodes
North Elevation of Teaching Building 1:300
I II III IV V VI

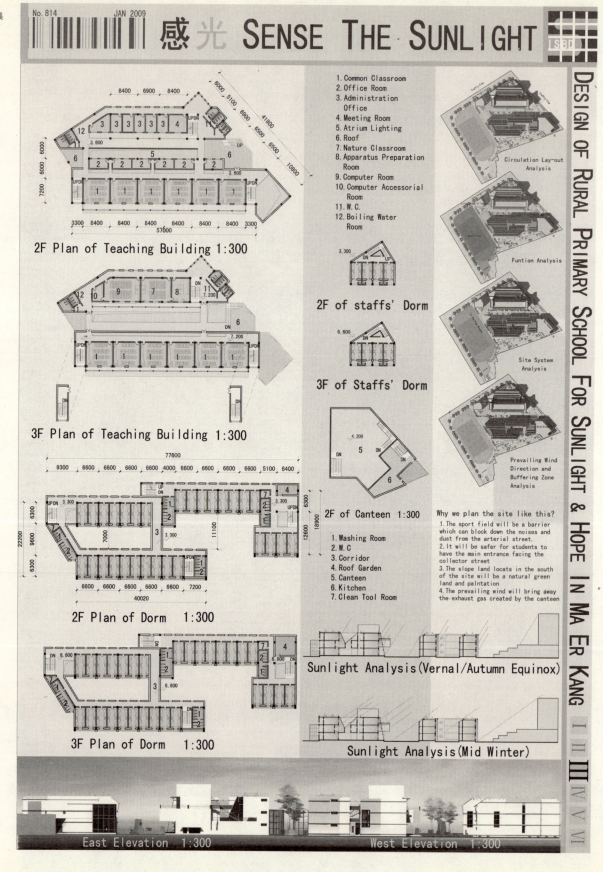

感光 SENSE THE SUNLIGHT

Design of Rural Primary School for Sunlight & Hope in Ma Er Kang

The design adopts simple retangular solid approach, which has a smaller shape coefficient. It is quakeproof and helps maintaining temperature of the building as well as saving energy.

In between the two rectangular solid buildings a common space with adequate thermal comfort, abundant natural light and nice view..

Apart from the outdoor playground, there is space on every floor of the teaching building. During winter times, these places are equiped with good thermal comfort and students can choose to play indoor.

When the distance between the two rectangular solid buildings suits the solar altitude in midwinter, it can satisfy the sunshine duration and save the lands.

As an angle of 35 degree is formed between the building and the base, a cuboid is inserted diagonally near the main entrance, parallel to the base. This is to ensure that no space is left between the building and the base, and also to make the building more powerful in appearance.

Culture of mountainous region Tibet and Qiang(羌) is integrated in the design of diagonally-inserted prism, forming a multi-layered and special main entrance.

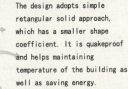
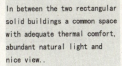

Geometry analysis

Teachers

Densely populated

Few people are

Distribution of population density

Students

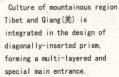
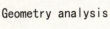
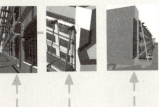

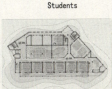
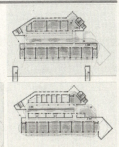

I II III **IV** V VI

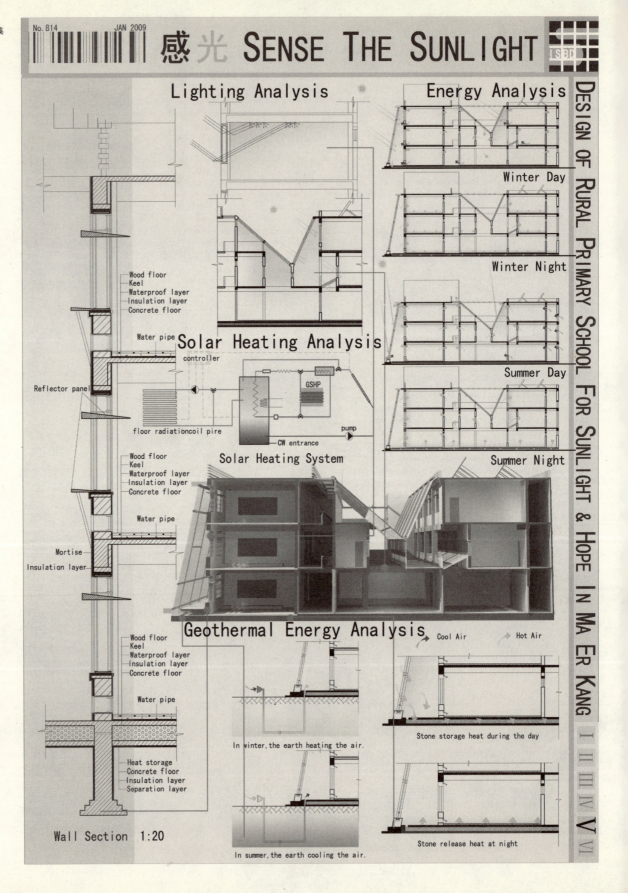

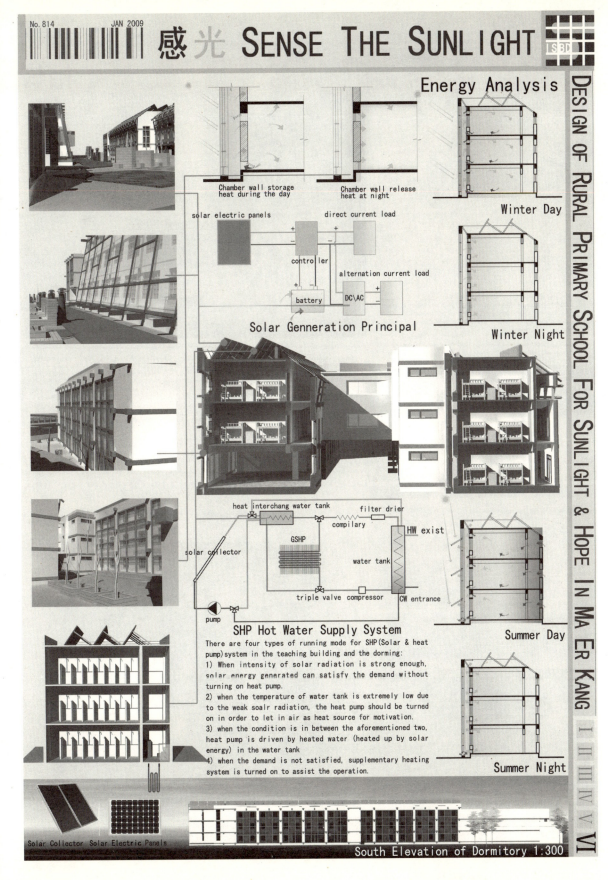

优秀奖
Honorable Mention Prize

项目名称：薪·火相传
Firewood Devolution

作　　者：刘骞、张乐、吉喆、甄密、韦今吾、郑一林

参赛单位：山东建筑大学建筑城规学院

薪·火相传 壹

石在，火种就在；
火种在，传承就不会停熄……

Stone in, fires on;
Fire in, passing out would not stop……

■ SUNSHINE ANLYSIS
To get better daylight, the terrain is the basis to all the houses builded.

■ WIND ANLYSIS
The main wind direction of Maerkang is WNW.
In the program, put polluted area, such as the school canteen kitchens and animal corner and so on, on the place under the direction.
West module and plants array locates to the north of town to prevent the cold wind in winter.
Considering proportion with the street, and coordinatething the best direction of wind and sunlight, these factors determine the direction of building.

To make full use of out-door space, architectural space environment will becombined with the free extension.

CULTURE — Residential areas in Sichuan with unique distinctive local characteristics.

SPACE

MATERIAL — Rational use of local building materials, such as stone and wood, to prevent the cold and wet.

本方案选址于马尔康，马尔康藏语意为"火苗旺盛的地方"。生活在这片优美土地上的孩子们正如寓意中的火种，孕育着希望，所以我们要像呵护火种那样，用心去关爱学生。方案采用了四川民居中的一些元素，在空间、尺度和材质上，让孩子们随时随地感觉亲切，营造了一种家的感觉。充分考虑当地太阳能资源情况，采用的技术以被动式太阳能为主，利用建筑结构本身的附加阳光间，直接受益窗和集热蓄热墙等，另外还适量运用太阳能真空管空气集热技术、太阳能低温辐射地板采暖技术和太阳能烟囱等主动式太阳能。通过上述技术措施，不仅能够保证室内的温度及空气的品质，而且能够节约大量的一次性能源和运行成本。

The base is chosen in Ma'er kang which is called "the place where the fire is flower" in Tibetan. Children who live on this piece of the beautiful land is as the meaning of fire, with hope. Therefore, we have to care them as fire. Some of the residential elements are used in the design. In space, scale and materials, let the kids feel very genial anytime, anywhere, to create a feeling of home.
Give full consideration to the local solar energy resources. Passive solar energy which is the main technology is used. The use of building structure itself like attached sunspaces, direct system, thermal storge wall and so on. Besides, vacuum tube solar collectors, low temperature hot water floor radiant heating and solar chimney are used as active solar energy technology. By the technology measures mentioned above, not only the temperature and the quality of the air can be kept stable, but also it can save a large number of one-time energy and operational costs.

LOCATION: MAERKANG, SICHUAN, CHINA

In order to create a feeling of home for children, program take the form of residential, such as sloping roof, the evolution of an open courtyard in the form, gallery across the street, decoration and so on.

Make full use of the natural energy, such as adequate lighting conditions, the impact of monsoon, fresh air, abundant rainfall, to take eco-technical measures, ventilation measures, rainwater harvesting.

For primary and secondary school students, physical and spiritual care are particularly important. Therefore, program in the scale and space Duli check should Implement "people-oriented."

Analysis of sunshine hours

CUTLINE(HOUR)
0
1
2
3
4
5
6

In the program base, there is a five-meter-high slope in the south. Taking into account the noise from the street, put the dorm along the slope in order to ensure the privacy of student hostels. On the slope, there is a five-storey building. In order to avoid the sun-shading of the dorm, a winter sunshine day analysis was done according to the height of slope, floor and windowsill. The results showed that: all quarters to meet the relevant norms of the minimum requirements of sunshine hours.

MASTER PLAN 1:500

SOLAR PRIMARY SCHOOL

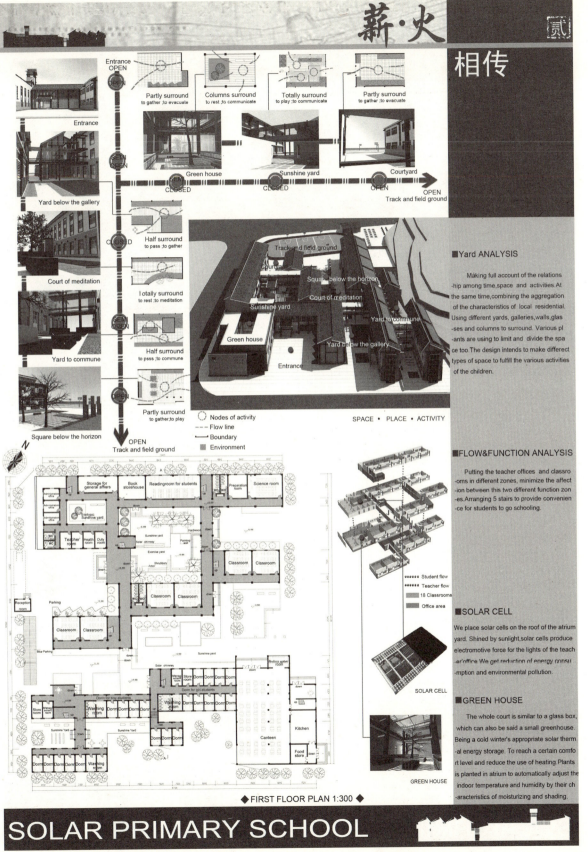

SOLAR PRIMARY SCHOOL

Yard ANALYSIS

Making full account of the relationship among time, space and activities. At the same time, combining the aggregation of the characteristics of local residential. Using different yards, galleries, walls, glasses and columns to surround. Various plants are using to limit and divide the space too. The design intends to make differect types of space to fulfill the various activities of the children.

FLOW&FUNCTION ANALYSIS

Putting the teacher offices and classrooms in different zones, minimize the affection between this two different function zones. Arranging 5 stairs to provide convenience for students to go schooling.

SOLAR CELL

We place solar cells on the roof of the atrium yard. Shined by sunlight, solar cells produce electromotive force for the lights of the teacher'office. We get reduction of energy consumption and environmental pollution.

GREEN HOUSE

The whole court is similar to a glass box, which can also be said a small greenhouse. Being a cold winter's appropriate solar thermal energy storage. To reach a certain comfort level and reduce the use of heating. Plants is planted in atrium to automatically adjust the indoor temperature and humidity by their characteristics of moisturizing and shading.

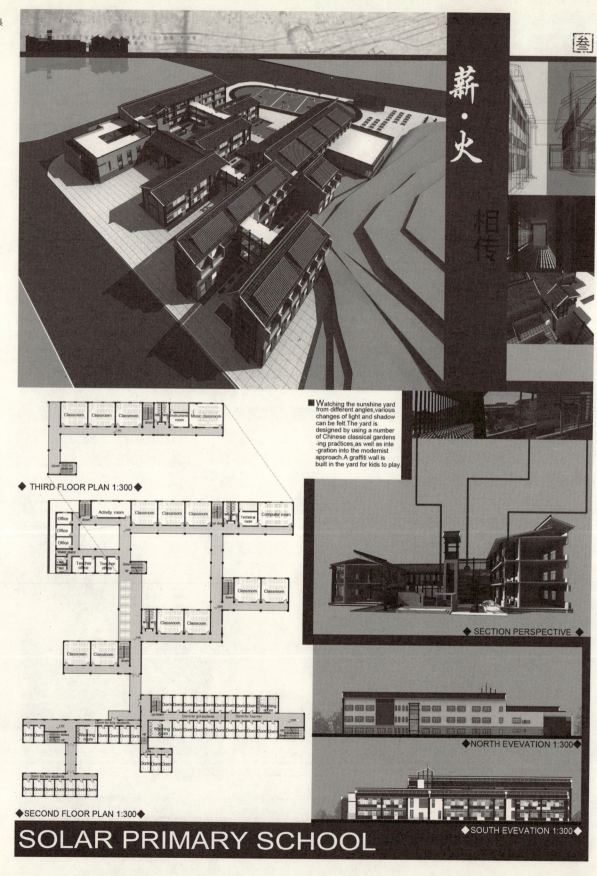

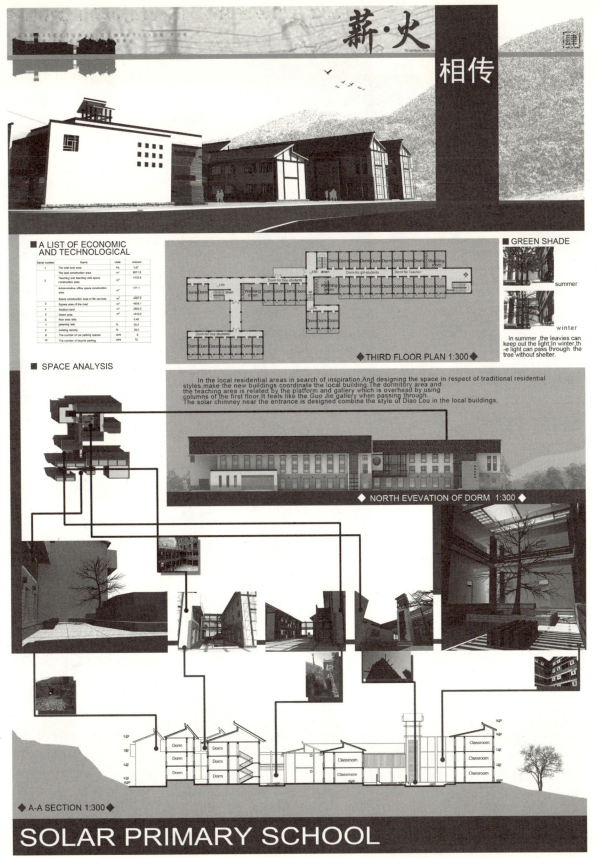

SOLAR PRIMARY SCHOOL
薪·火相传

Overview
Design of ecological architecture doesn't necessarily have to rely on the latest high-tech means. The ecological architecture do not necessarily have to cost a lot. We must not only conside the future of economic and environment, but also the deeper social significance about the society. The most important, bringing about sustainable develop-ment for the future generations is the ultimate goal.
The main wind direction of Maer-kang is WNW. We analysis and research the local use of solar and the analysis available to the following two analysises.

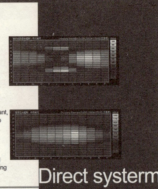

Low temperaure hot water floor radiant heating
■ Solar water heating is placed on the sloping roof of the hostel and supply hot water for radiant heating sysyterm to improve indoor comfort level of the hostel.

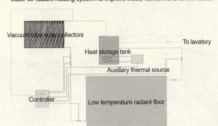

solar hot water floor radian heating

Floor section plan

Thermal storage wall
■ The adobe wall is used to insulate. Adobe wall shined by sunlight as soon as the sun rises will be warm soon and ensure a comfortable internal environment that all the students will be able to enjoy in the cold winter morning.

Direct systerm
Some quarters of hostel benefit directly from the use of South window. In the cold winter direct sunlight will rapidly increase the internal temperature.

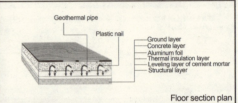

Floor panel heating
■ Attached sunspace is combined with the collector. Through lots of theoretical analysis and practical tests, we get results that can be attributed to the following two charts.

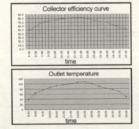

Technology
■ According to the careful analysis of the local ventilatin, sunshine and other conditons we find that some problems such as shade, heating, ventilation and so on.

Large samples

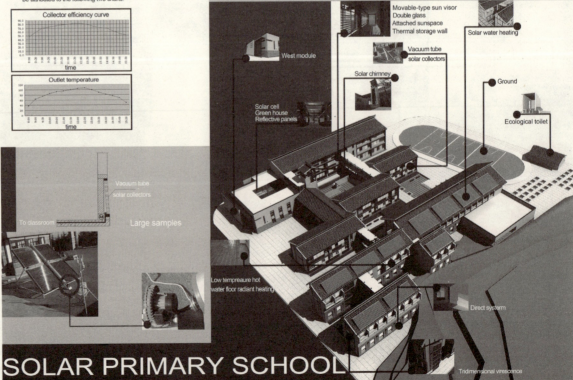

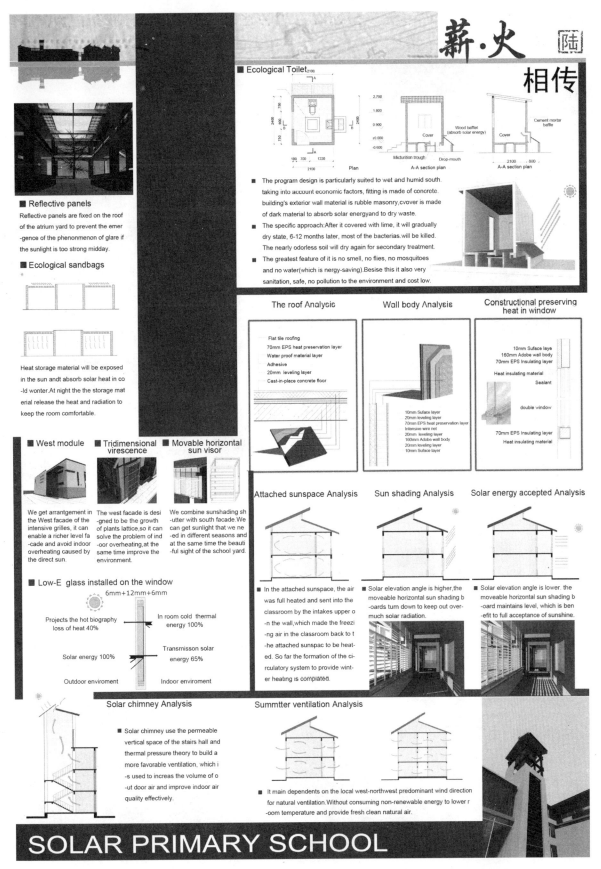

优秀奖
Honorable Mention Prize

项目名称：巢
　　　　　Nest
作　　者：赵晓娜、刘铮、韩超、马卿、
　　　　　白叶飞、聂广强、尚大伟
参赛单位：内蒙古工业大学

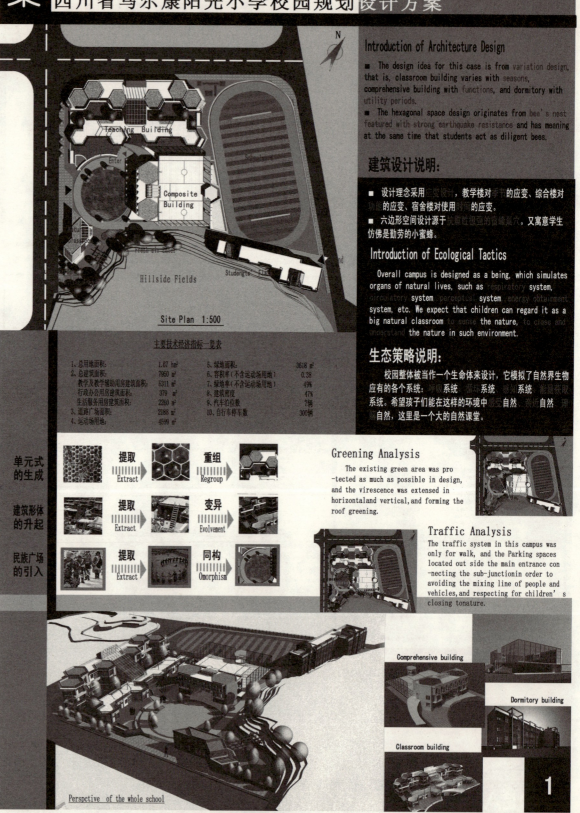

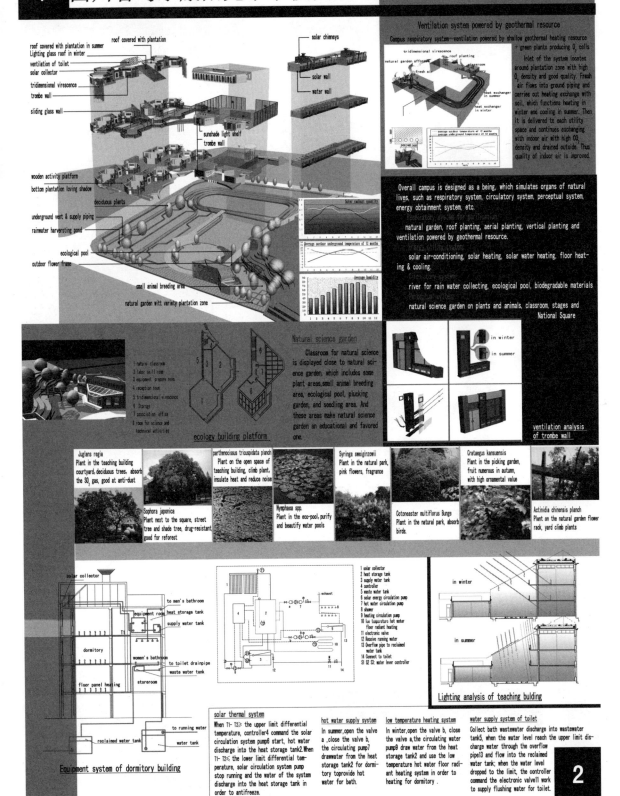

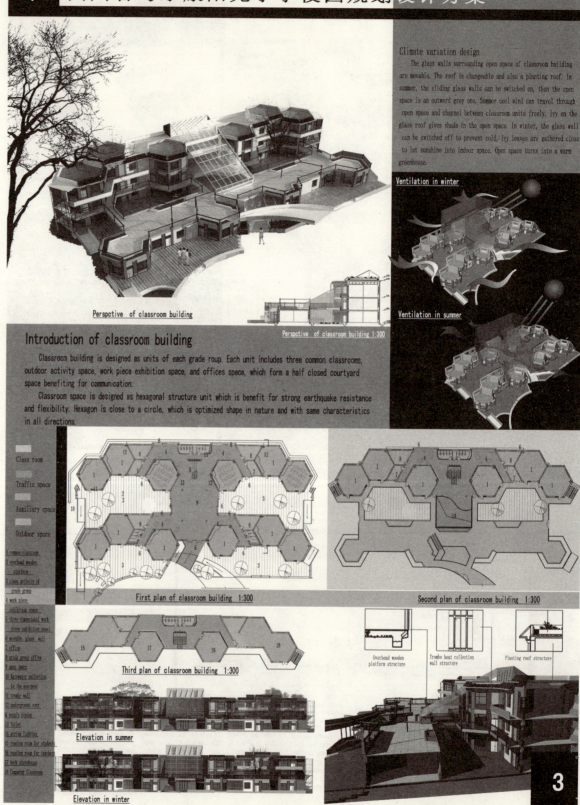

四川省马尔康阳光小学校园规划设计方案

959号 巢

2009台达杯国际太阳能建筑设计竞赛获奖作品集

Introduction of Comprehensive Building

Comprehensive building has realized variation space design. Roof of dining hall can be acted as course for basketball and volleyball. Dining hall can be as dancing room or outdoor playground during no-meal time. Music classroom can be changed into a stage for performance activity after the sliding walls are moved away. High platform entrance to comprehensive building can act as outdoor stage for national squares or performance stage for national opera stage for villages.

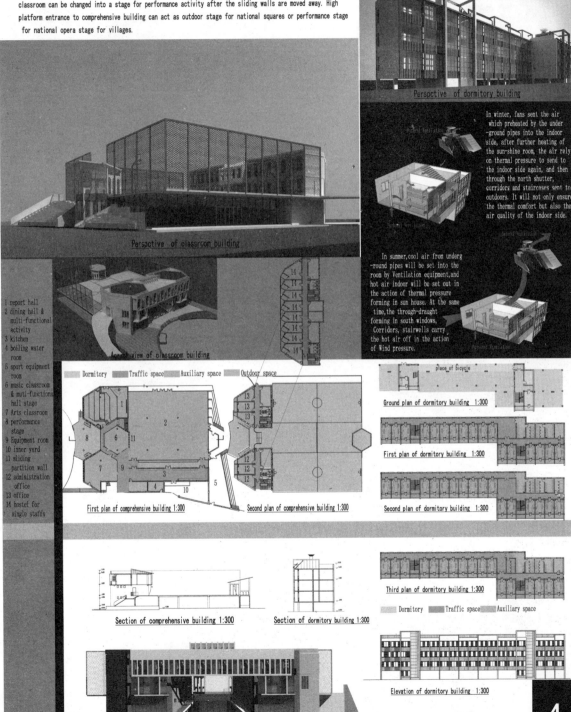

In winter, fans sent the air which preheated by the underground pipes into the indoor side, after further heating of the sun-shine room, the air rely on thermal pressure to send to the indoor side again, and then through the north shutter, corridors and staircases sent to outdoors. It will not only ensure the thermal comfort but also the air quality of the indoor side.

In summer, cool air from underground pipes will be set into the room by Ventilation equipment, and hot air indoor will be set out in the action of thermal pressure forming in sun house. At the same time, the through-draught forming in south windows, Corridors, stairwells carry the hot air off in the action of Wind pressure.

1 report hall
2 dining hall & multi-functional activity
3 kitchen
4 boiling water room
5 sport equipment room
6 music classroom & muti-functional hall stage
7 Arts classroom
8 performance stage
9 Equipment room
10 inner yard
11 sliding partition wall
12 administration office
13 office
14 hostel for single staffs

4

二、技术专项奖作品
Prize for Technical Excellence Works

NO. 1

SUNSHINE & HOPE

技术专项奖
Prize for Technical Excellence

项目名称：阳光与希望
　　　　　Sunshine & Hope
作　　者：Brian Messana、Toby O'Rorke、
　　　　　Sungpyo Kim、Christopher Biggin
参赛单位：Messana O'Rorke Architects

专家点评：
作品很有趣，很有创新的设计。在绿色屋顶下提供了所有的需求空间。屋顶内空闲空间使用较好，太阳辐射可以有效地进入屋内以提供自然采光和被动采暖。但是能量概念不清楚，对场地退线的利用以及建筑消防等方面不具有可操作性，采光和自然通风实施起来也有难度。

The works is interesting and it is such a design with originality. All of spaces needed are provided under a green roof. The spare space within the roof is good in use and solar radiation may effectively come into the interior for natural light and passive heating. However the concept of energy is not clear enough. It is not operable in aspect of fully utilization of the field, building fire protection, etc. It is also difficult in implementation of lighting and natural ventilation.

Inspired by the agrarian landscape of China and the haphazard plan of a medieval Chinese village, we sort to create a self sustaining living and working environment.
The entire school and living facilities exist below a single green roof, which is a landscape of natural vegetation, gardens, play areas and vegetable plots. Punctured through this surface are courtyards of various sizes, based on an abstraction of the plan of Taoping. On the perimeter of these courtyards the various functions of the facility are accommodated. Each of the four dormitory "houses" are based around private courtyards, with two stories of sleeping and washing facilities for children and staff, and each include a common area in which the children of that house can interact. The communal canteen looks over the outdoor multi-use sports field which occupies the large central court around which the classroom and school facilities are arranged. Conceptually the scheme with its massive exterior walls and central courtyard mirrors traditional Chinese architecture and provides the same safe nurturing environment. This scheme is designed for the Mian Yang site, but the concept could be applied to any site and adjusted according to program.

MAIN TECHNICAL AND ECONOMIC INDICATORS

1	Total Site Area:	13,450 Sq M
2	Total Floor Area:	7,532 Sq M
	Floor Area of Teaching and Accessory Rooms:	3,158 Sq M
	Floor Area of Administration:	350 Sq M
	Floor Area of Service:	850 Sq M
3	Road and Paved Area:	671 Sq M
4	Sports Field:	3862 Sq M
5	Green Area:	5346 Sq M
6	Floor Area Ratio (excluding sports field):	56%
7	Green Area Ratio (excluding sports field):	39%
8	Building Density:	56%
9	Car Parking Spaces	4
10	Bike Parking Spaces	750

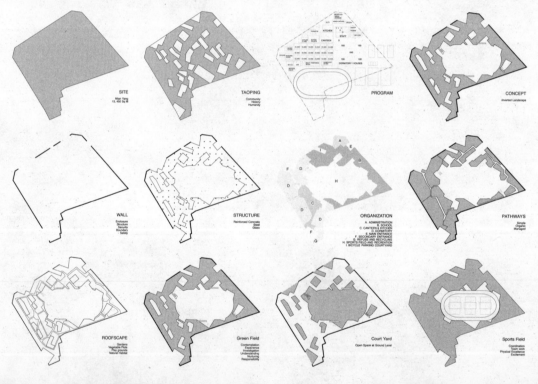

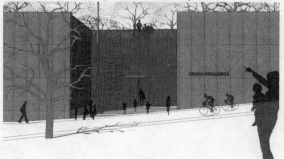

Perspective no. 1: looking towards school entry

Perspective no. 2: looking along shool towards dormitory entry

NO. 2

Sustainable Design

Green Roof - Green Field

The Green Roof acts as a water collection and filtration method for the building. The collected water can be either instantly used as grey water or cleaned and then used as drinking water.

Vegetable gardens are situated on the roof and are both maintained by the students and used to supplement food supplies.

Vegetable waste from the kitchen is composted and then subsequently used in the gardens.

This green layer also acts as a thermal blanket.

Solar Systems

Passive solar gain through massive exterior concrete walls and roof - absorb heat during the day and release at night

Large areas glass orientated for maximum solar gain.

Passive solar water heating to augment hot water resources.

Courtyards

The courtyards act as lightwells allowing light to penetrate the inner sections of the building

These spaces allow also for natural ventilation and solar heating in the buildings

Seasonal Shading

The trees offer shade in the summer and then light to penetrate through to the building in the winter when they loose their leaves

Geothermal Heat Pump

A geothermal pump is installed under the building. This is used to regulate the temperature of the building.

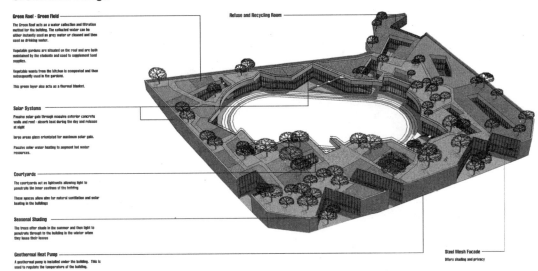

West Elevation

Perspective no. 3: looking towards dormitory entry

Perspective no. 4: looking down towards school

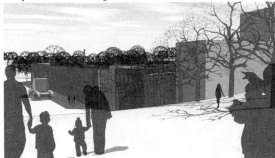

NO. 3

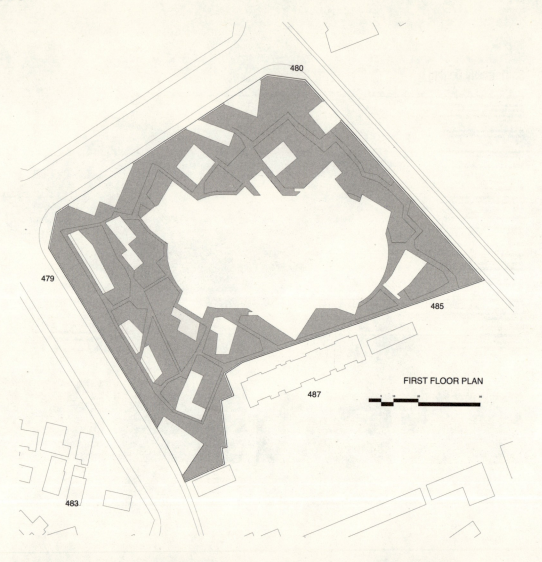

FIRST FLOOR PLAN

North Elevation

Perspective no. 5: main entrance looking toward athletic field

Perspective no. 6: looking towards main school entry

NO. 4

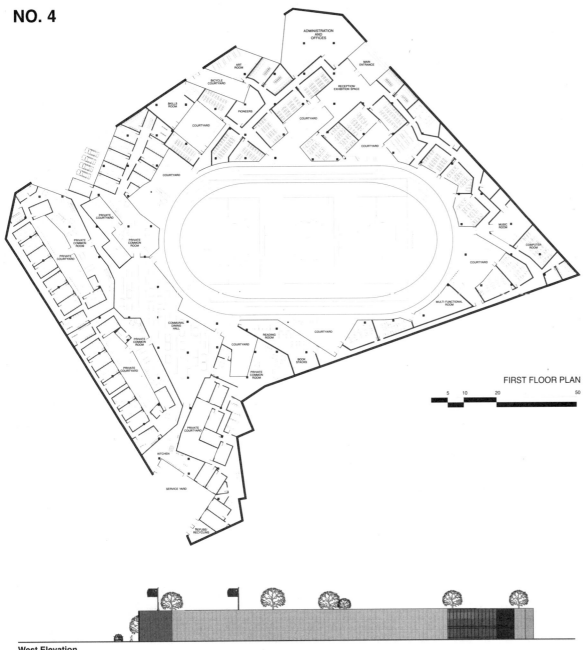

FIRST FLOOR PLAN

West Elevation

Perspective no. 7: view of athletic field

Perspective no. 8: athletic field looking at stair to roof garden

NO. 5

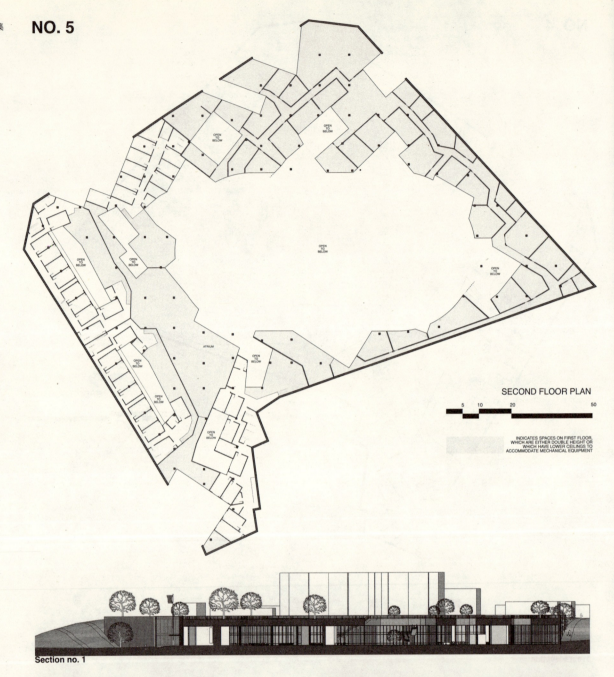

SECOND FLOOR PLAN

5 10 20   50

INDICATES SPACES ON FIRST FLOOR,
WHICH ARE EITHER DOUBLE HEIGHT OR
WHICH HAVE LOWER CEILINGS TO
ACCOMMODATE MECHANICAL EQUIPMENT

Section no. 1

Perspective no. 9: view of roof garden

Perspective no. 10: looking down from roof garden to court yard

NO. 6

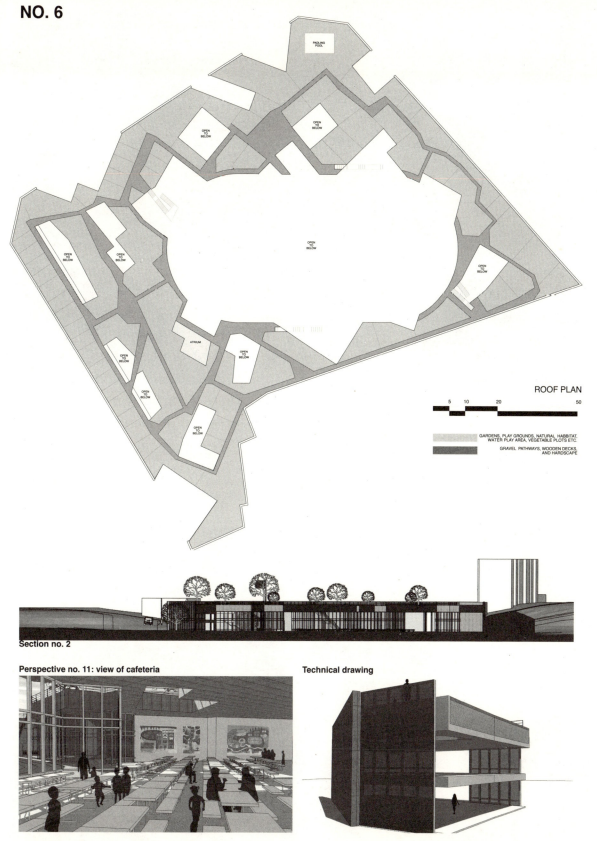

技术专项奖
Prize for Technical Excellence

项目名称：太阳伞
　　　　　Solar Umbrella

作　　者：杨鹏程、黄一滔、李荣、吕家悦、
　　　　　何兆熊、谢崇实、冷冷

参赛单位：重庆大学建筑城规学院

专家点评：

分析详细，细部设计较好。建筑造型来源于"竹筒"和"谷堆"这两种乡土味的形式，配合"可变组合表皮"，对遮阳、通风等形式做了整合设计，很有创新。缺点是结构形式比较复杂，不利于实施。

The scheme has an analysis in detail and a better detail design. The idea of architectural appearance comes from the forms of "bamboo tube" and "corn pile", which are full of native smell. Combining with "alterable and combinatorial cuticle" there is something original with integrated design regarding the forms of sun shade and ventilation. The shortcoming is that the structural type is complicated and not profitable to the construction.

362 四川省绵阳阳光小学设计

策略分析与形式的整合　太阳伞

BIRDS'EYS VIEW

设计说明

方案构思源自乡村生活带来的灵感，立足于采用乡土材料，建构具有乡土气息的建筑，同时充分利用太阳能等可再生能源进行绿色建筑设计。选取"竹"作为主要的建筑材料，建筑造型脱胎于"竹筒"和"谷堆"这两种乡土味的形式。利用竹编织、搭接等方式形成多种以应对不同方位和气候条件的"可变组合表皮"。根据灾区日照、太阳辐射、风等生态因素，采用多种主被动式太阳能及其他技术，进行建筑一体化设计，通过大胆、富有创造性的设计，为灾后重建提供了一种新的建筑形式。

Designing conception is derived from the rural life of inspiration. Based on the use of local materials, Construct the building with a local flavor, At the same time, make full use of solar and other renewable energy sources for green building design. Select the "Bamboo" as the main construction materials, Architectural modeling derived from two local flavor of" bamboo tube " and " grain mass ". The use of bamboo weaving and lapping, to deal with a variety of different orientations and climatic conditions, to form"variable combination skin." According to the sunshine, solar radiation, wind and other ecological factors of disaster areas, Using a variety of active and passive solar energy and other technical to proceed integrated design, Through bold, creative design provide a new architectural forms for post-disaster reconstruction.

SKYLIGHT LEVEL ANALYSIS

VENUES RADIATION ANALYSIS

SITE PLAN 1:500

策略分析与形式的整合 太阳伞

设计中根据光、太阳辐射、自然风、噪声的因素对建筑的影响,将外表皮分为东南西北四个区,通过不同的可变的构造方式,对遮阳,通风,太阳辐射以及采光做针对性的处理。
According to light, solar radiation, natural wind, noise factors on the impact of construction,the skin be divided into four areas,through variable structure by means of shading, ventilation, solar radiation, as well as do lighting for of treatment.

WEST STRATEGY: Double Glazing and level shutter compose the vertical sun visor, To solve the summer glare and overheating problems.

NORTHERN STRATEGY: To adopt the central axis of rotating windows, through the rotation combination ,to solve the summer issue of natural ventilation and thermal insulation in winter.

 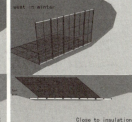 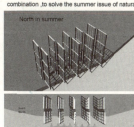 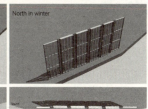

Summer: Sun visors turn into vertical shade after opening vertical. According to the summer wind-led ,adjust the angle of sun visor to lead Natural wind into Indoor.

Winter: Rotary Close the Sun visor to keep warm.

Summer: The windows open, to facilitate import of the summer monsoon.

Winter: Rotating the windows closed, after the mutual occlusion ,to Composed of double-envelope structure.and strengthen the construction insulation during the winter.

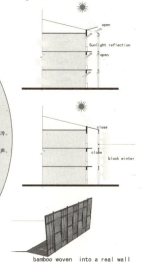

SOUTHERN STRATEGY: The use of combination of double facade windows and level shutter windows, Through the central axis of rotation to resolve the summer shading, ventilation and heating problems in winter.

  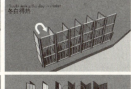 

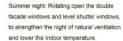  

Summer night: Rotating open the double facade windows and level shutter windows, to strengthen the night of natural ventilation and lower the Indoor temperature.

Summer daytime: Rotating open the double facade windows of the combination window, At the same time, rotating close the level shutter windows to form the level shading,in order to block the Strong sunlight in summer.

Winter daytime: Rotating open the level shutter windows of the combination window, At the same time, rotating close the double facade windows. To enable the outside balcony formation of the solar house of direct access to the heat.

Winter night: Rotating close the double facade windows and level shutter windows. Composed of double-envelope structure, and strengthen the construction insulation during the night time.

362 四川省绵阳阳光小学设计

策略分析与形式的整合 太阳伞

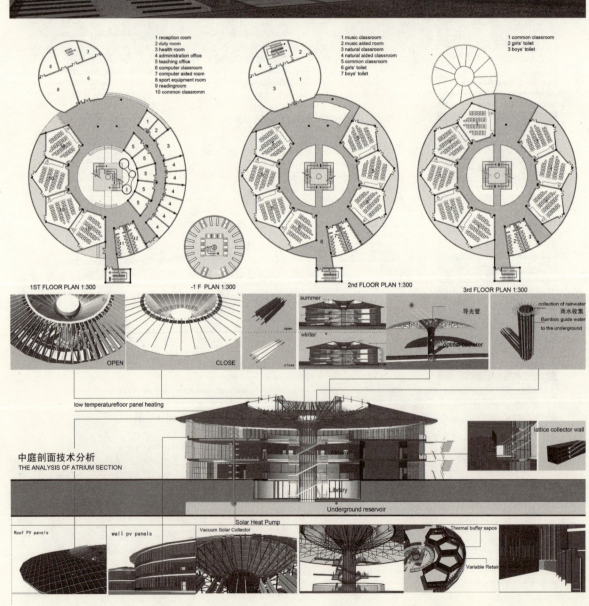

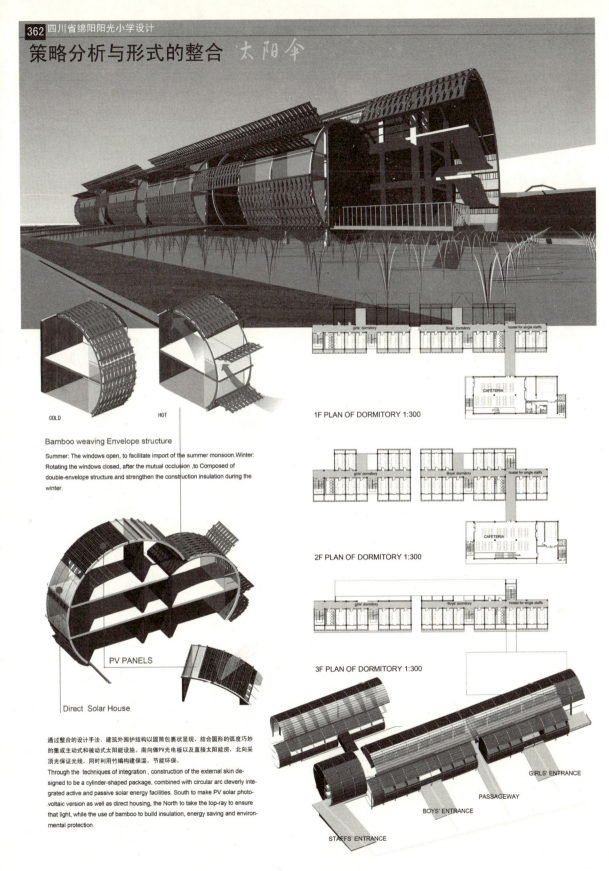

技术专项奖
Prize for Technical Excellence

项目名称：绵阳阳光小学
The Sunshine Primary School in Mianyang
作　　者：吴忠勋
参赛单位：台湾交通大学建筑研究所

专家点评：

"社会与环境互动的小学"概念很新颖，自然采光、自然通风和雨水收集的原理分析很清晰。但建筑结构过于复杂，不利于把握被动设计的整体效果。

"A social and environmental interactive primary school" is a new conception. The analysis concerning natural lighting, natural ventilation and rain water collection is very clear. However the structure of buildings is too complicated to hold integrated effect of passive solar design.

CONCEPT AND STRATEGY

· 基地位置：绵阳

· 永续的校园
由于震灾后的地区需要大量的小学重建，主要设计概念为建立一个能 与社会与环境互动的小学建筑，透过互动使大量小学的建设、维持更为容易，避免过于快速的建设而无法永续维持。

· 阳光小学＝太阳能发电厂＋社区公园
透过与官方电力单位的合作，共同建立大量的太阳能板于小学中，发电供给校园使用，多余电力由电力单位回收。如此可降低小学造价，并供农村学童更好的学习及住宿的环境。
社区公园的设立则能增加学童与社区的互动，并供给社区新的活动空间。

· **Site：Mian-Yang**

· **Sustainable primary school**
Because of the earthquake disaster, many areas need the massive reconstruction of primary school. Mainly designs concept is to establish the school construction system that can be interacted with the society and environment. It makes the construction and maintenance could be easier by the interaction, and avoid the fast construction being unable to maintain.

· **Sunshine primary school = Solar power plant + Communal park**
Though the cooperation with the official electric power organization, establishes the massive photovoltaic system in the primary school, the electricity generation supplies campus use, the surplus electric power recycles by the electric power organization. So it may reduce the cost of school construction, and supplies the schoolchild from countryside a better environment to study and lodge.
The communal park can increase the interaction between schoolchild and community, and supplies the community new activity space.

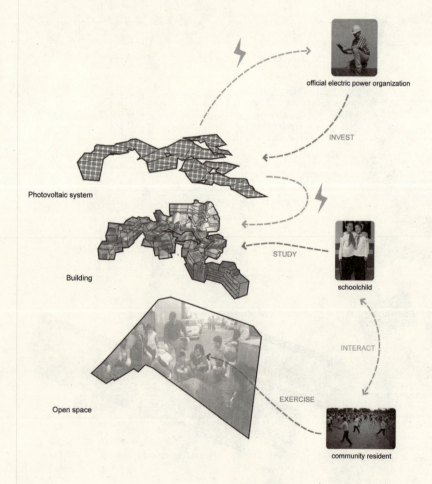

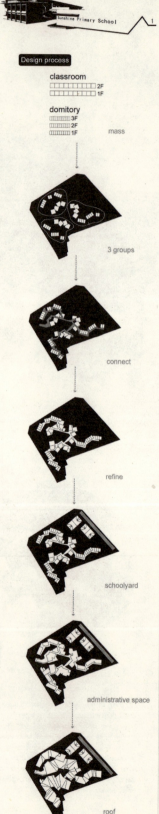

GROUND FLOOR PLAN 1/500

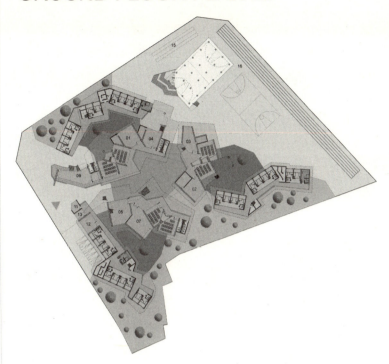

2nd FLOOR PLAN 1/500

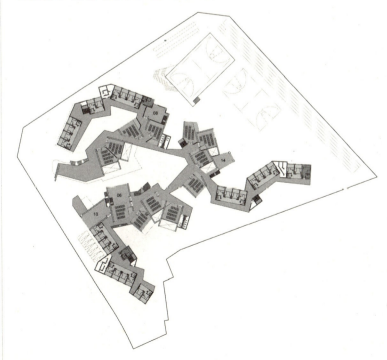

SITE PLAN

In order to make the most effective electricity generate, the roofs are covered with BIPV. The area of BIPV about 5020m², almost 30% of the site.

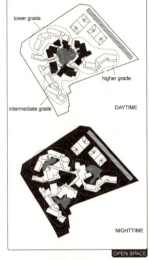

OPEN SPACE

The buildings are divided into 3 groups according to the age of students. Each group has its own open space, and the 3 groups are connected by deck on 2nd floor.

▶ MAIN ENTRANCE

■ COMMON CLASSROOM

■ PARTICULAR COURSE CLASSROOM
01 music classroom
02 natural science classroom
03 computer classroom
04 painting classroom
05 reading room for students
06 book storehouse
07 multi-purpose classroom
08 labor skill room

■ ADMINISTRATIVE ROOM
09 administration office
10 teacher office
11 duty room
12 storage for general affairs
13 health care room
14 association office

■ HOSTEL FOR STAFFS

■ DORMITORY

□ CANTEEN

15 BIKE PARKING

16 SCHOOLYARD

ID : 457

3rd FLOOR PLAN 1/500

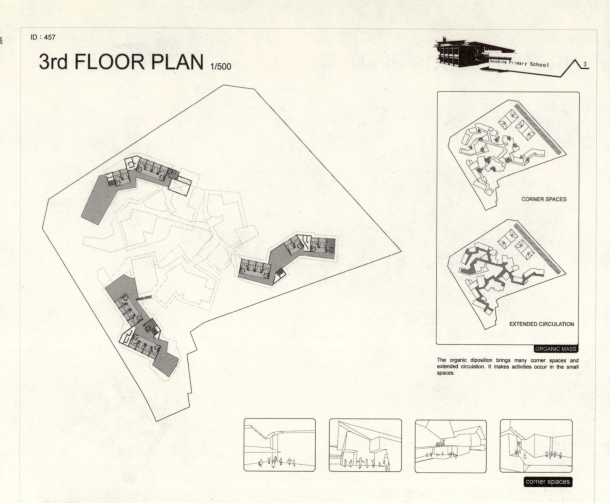

CORNER SPACES

EXTENDED CIRCULATION

ORGANIC MASS

The organic diposition brings many corner spaces and extended circulation. It makes activities occur in the small spaces.

corner spaces

TRADITION AND TECHNOLOGY

The "through type timber frame (chuandou framing system)" is a kind of structure type of Sichuan traditional houses. This kind of structure support the house by load-bearing wall which construct by combined pillars. The building in this case transfer this kind of structure into new structure form.

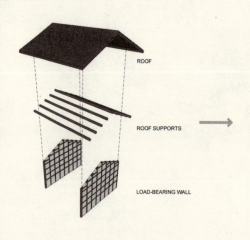

ROOF

ROOF SUPPORTS

LOAD-BEARING WALL

Traditional "through type timber frame"

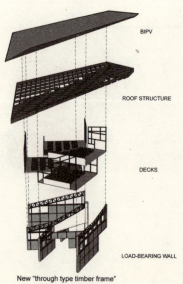

BIPV

ROOF STRUCTURE

DECKS

LOAD-BEARING WALL

New "through type timber frame"

SECTION 1/100

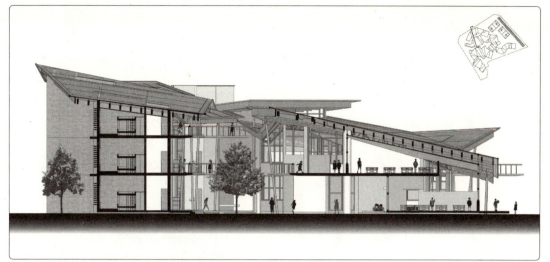

SOLAR TECHNOLOGY DOUBLE ROOF

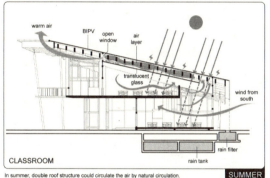

In summer, double roof structure could circulate the air by natural circulation. The translucent glass can make ambient light into the classroom.

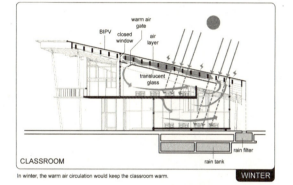

In winter, the warm air circulation would keep the classroom warm.

PERSPECTIVE

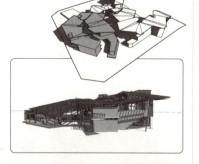

技术专项奖
Prize for Technical Excellence

项目名称：家·园·院
My Family & My School & My Paradise
作　者：王天晖、徐斌、丁瑜
参赛单位：东南大学建筑学院

专家点评：

流线型的建筑组织形式很清晰，外观整齐，内部功能分区合理，能量系统分析比较到位，尤其是景观和人工湿地的设计很有特点。缺点是概念分析比较多，可操作性不强。

In this scheme the organization of clipper-built buildings is clear and it has an orderly appearance. Interior functional zoning is reasonable. It has a good analysis about energy system. Especially the design of sight and man-made marsh is remarkable. The shortcoming is more of concept analysis and less of operability.

MY FAMILY & MY SCHOOL & MY PARADISE
家·园·院

建筑构思：塑造一个整体的建筑形象，蕴含着一个家庭和一个学校的寓意。学校既是学生学习的地方，同时又是灾区学生的家园，更重要的是他们的精神家园。形成了"家园院"的空间模式，家——家庭，园——校园，院——庭院，形成了生活、学习、活动的综合性校园。适合灾区的实际情况，具有一定的现实意义。

Architecture concept: to shape a building, which contains a the moral of family and school. School is a place where students learn, but it is also home to the students in disaster area, more important, it is their spiritual home. Formed a "Home - Park - Garden" space model, home - the family, park - the campus, garden - the courtyard. It forms a living, learning, sporting - a comprehensive campus. It adapts to the actual situation of the disaster area and has a certain practical significance.

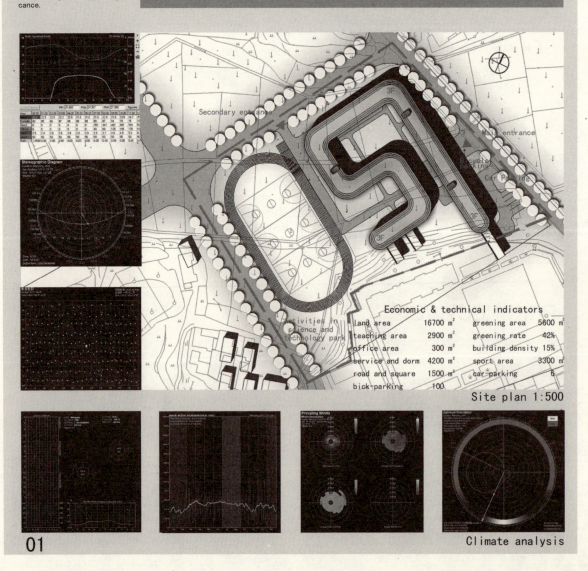

Economic & technical indicators

land area	16700 m²	greening area	5600 m²
teaching area	2900 m²	greening rate	42%
office area	300 m²	building density	15%
service and dorm	4200 m²	sport area	3300 m²
road and square	1500 m²	car-parking	6
bick-parking	100		

Site plan 1:500

01　　Climate analysis

MY FAMILY & MY SCHOOL & MY PARADISE

建筑材料：双层表皮，外层运用当地竹材，形成可推拉移动的表皮，适应季节和昼夜的变化，夏季可以围合出外廊空间，冬季可作为阳光室来使用。同时外表皮上挂太阳能板来收集太阳能加以利用。
Building materials: double-surface, the outer layer is bamboo, which can form flexible surface, in order to adapt to the alternating of seasons and day / night. In summer, it can forms gallery-space, in winter it can be used as a sunshine room. At the same time solar panels can be hanged on surface to collect solar energy.

MAR. 22 10AM

MAY. 22 10AM

JUN. 22 10AM

OCT. 22 10AM

DEC. 22 10AM

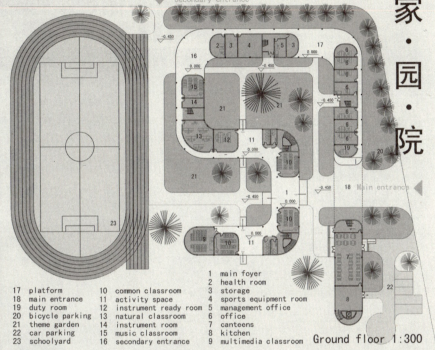

家·园·院

Ground floor 1:300

1 main foyer
2 health room
3 storage
4 sports equipment room
5 management office
6 office
7 canteens
8 kitchen
9 multimedia classroom
10 common classroom
11 activity space
12 instrument ready room
13 natural classroom
14 instrument room
15 music classroom
16 secondary entrance
17 platform
18 main entrance
19 duty room
20 bicycle parking
21 theme garden
22 car parking
23 schoolyard

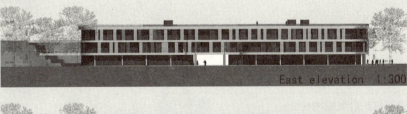
East elevation 1:300

West elevation 1:300

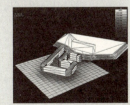

Sunlight analysis

MY FAMILY & MY SCHOOL & MY PARADISE

建筑空间形体：建筑采用流线型，在外观上是个整体，内部功能分区明确，整个建筑塑造了多处庭院与室内灰空间，为学生创造了不同的活动场地，此外建筑顶部的空间可以作为屋顶活动平台，提供了多层次的活动空间，另外建筑有桥梁连接到南侧的山上，化消极因素为积极因素，把场地的各个元素很好的结合在一起。

Building's space and shape: the building is streamlined and whole externals, its function is clear internals. The building created a number of indoor courtyard and gray space for students, in addition, the top of the building can be used as the platform for activities, it provides a multi-level space. The bridge connect the building to the south side of hill, which turns negative factors into positive factors and joins various elements in the site together.

家·园·院

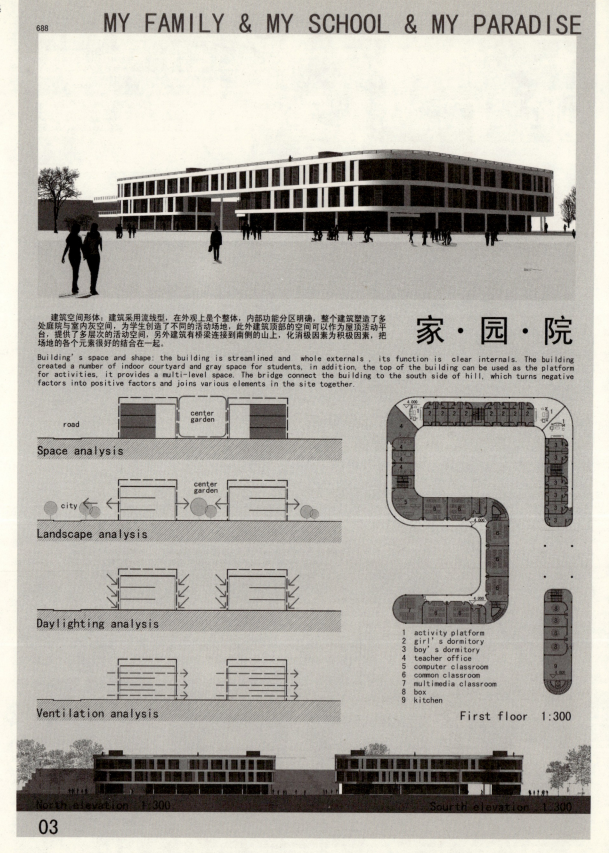

Space analysis
Landscape analysis
Daylighting analysis
Ventilation analysis

1. activity platform
2. girl's dormitory
3. boy's dormitory
4. teacher office
5. computer classroom
6. common classroom
7. multimedia classroom
8. box
9. kitchen

First floor 1:300

North elevation 1:300   Sourth elevation 1:300

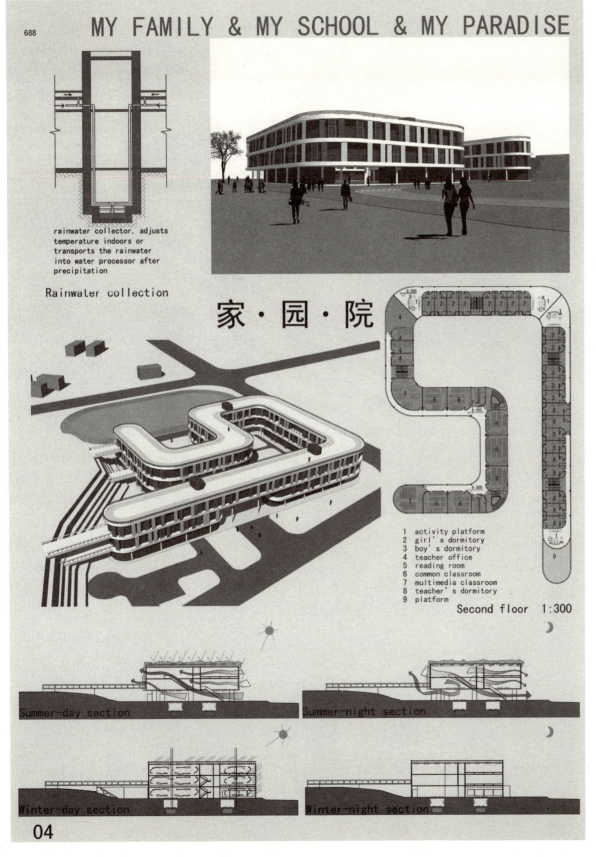

MY FAMILY & MY SCHOOL & MY PARADISE
家·园·院

Eco-energy saving technology

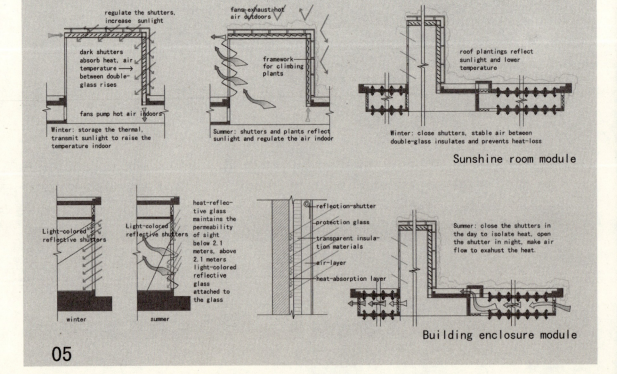

Sunshine room module

Building enclosure module

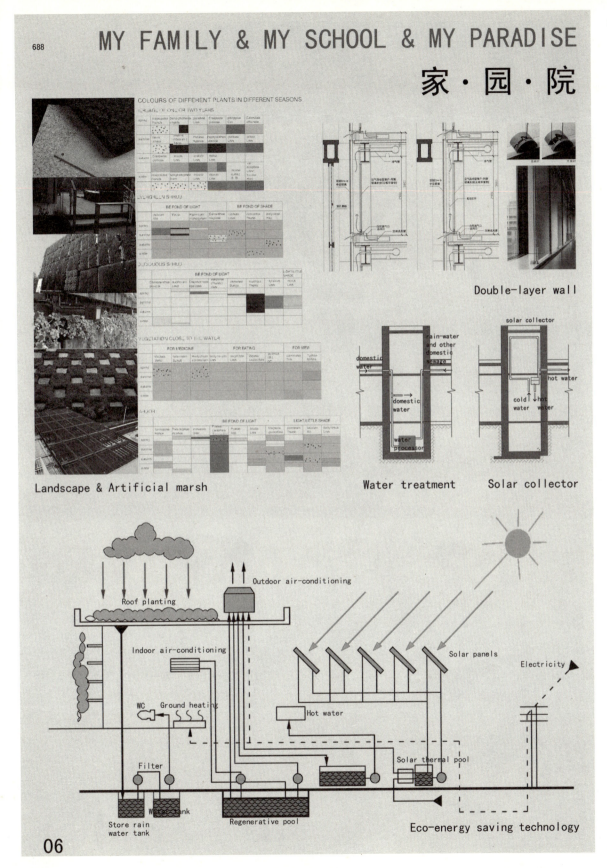

技术专项奖
Prize for Technical Excellence

项目名称：集合标准建筑
　　　　　Assembled-Standard Construction
作　者：贾雪子、刘芸
参赛单位：清华大学建筑学院

专家点评：
规划布局合理，根据教学和生活分为两个区域，每个区域各自独立。创新点在于采用高低错落的建筑组合形式解决了遮挡问题，并利用封闭中庭，增大的建筑室内活动空间、被动太阳房和太阳集热面积，是地处寒冷地区的马儿康地区利用太阳能解决采暖问题的较好形式。缺点是房屋布置比较单调，体量较大，不符合小学校的建筑特征。

The layout of the scheme is reasonable that the field is divided into two areas, teaching and living, which are independent to each other. The originality is that the arrangement of buildings with different heights at random is adopted, thus preventing sun shine from being sheltered by each other and the close atrium is used to expand the space for interior activities. Passive solar room and enough area of solar collection are good way to solve heating problem in cold Ma Er Kang district. The shortcoming is that the building arrangement is monotonous and the figure of the building is too big, which does not accord with the characteristics of a primary school.

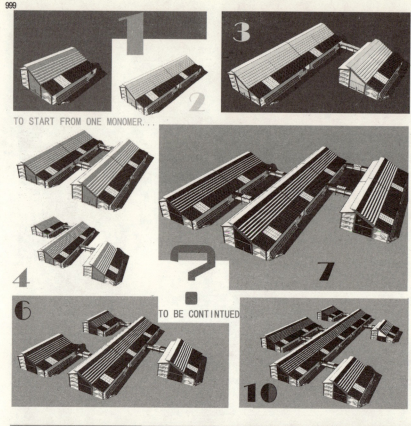

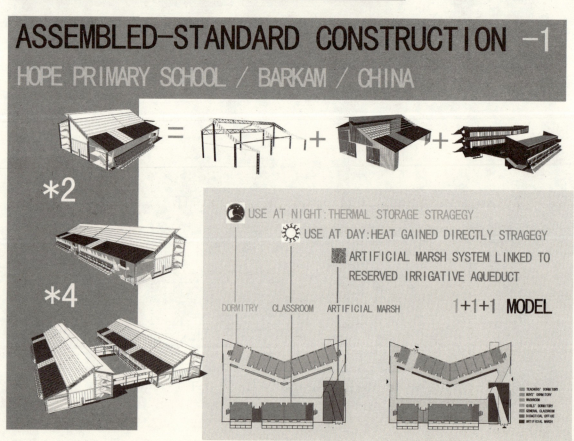

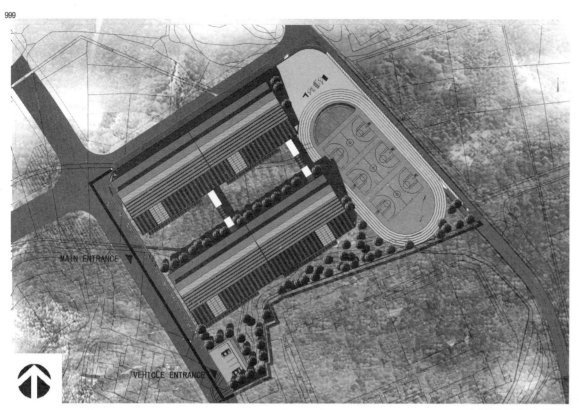

SITE PLAN 1:500

ASSEMBLED-STANDARD CONSTRUCTION -2
HOPE PRIMARY SCHOOL / BARKAM / CHINA

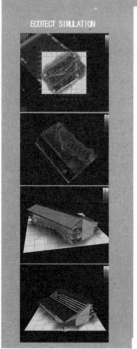

ECOTECT SIMULATION

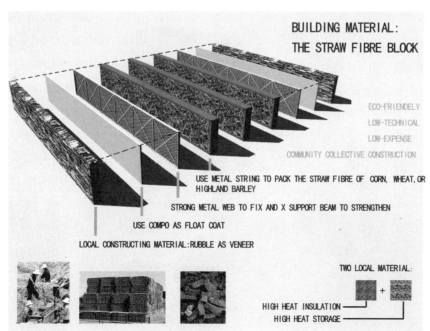

BUILDING MATERIAL:
THE STRAW FIBRE BLOCK

ECO-FRIENDELY
LOW-TECHNICAL
LOW-EXPENSE
COMMUNITY COLLECTIVE CONSTRUCTION

USE METAL STRING TO PACK THE STRAW FIBRE OF CORN, WHEAT, OR HIGHLAND BARLEY
STRONG METAL WEB TO FIX AND X SUPPORT BEAM TO STRENGTHEN
USE COMPO AS FLOAT COAT
LOCAL CONSTRUCTING MATERIAL: RUBBLE AS VENEER

TWO LOCAL MATERIAL:

HIGH HEAT INSULATION
HIGH HEAT STORAGE

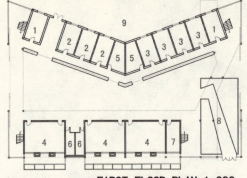
FIRST FLOOR PLAN 1:200

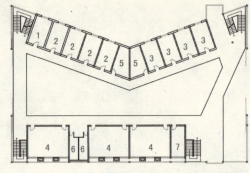
SECOND FLOOR PLAN 1:200

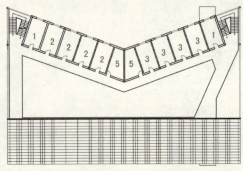
THIRD FLOOR PLAN 1:200

1- hostel for single staffs
2- boys' dormitory
3- girls' dormitory
4- common classroom
5- washroom & bathroom
6- washroom
7- teacher office
8- bridge over artificial marsh
9- bike parking

ASSEMBLED-STANDARD CONSTRUCTION -3
HOPE PRIMARY SCHOOL / BARKAM / CHINA

SUMMER MODEL
TWO HOUSE + ONE STREET
GATE OPEN
ROOF SHUTTER OPEN
SOLAR HOUSE OPEN
IVY BESTREW WALLS AND BALLUSTERS FLOURISH

WINTER MODEL
ONE WHOLE HOUSE
GATE CLOSED
ROOF SHUTTER CLOSED
SOLAR HOUSE CLOSED
IVY BESTREW WALLS AND BALLUSTERS SEAR

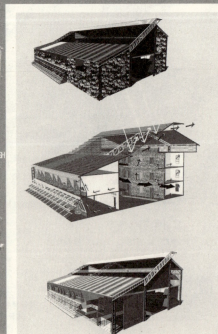
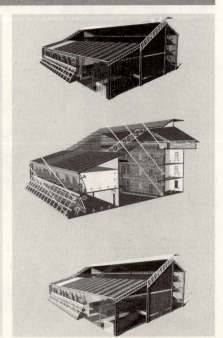

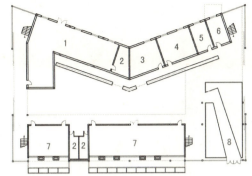
FUNCTIONAL UNIT FIRST FLOORPLAN 1:200

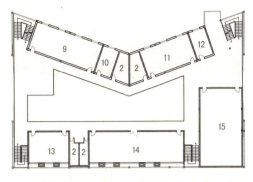
FUNCTIONAL UNIT SECOND FLOORPLAN 1:200

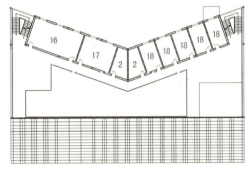
FUNCTIONAL UNIT THIRD FLOORPLAN 1:200

1- kitchen
2- washroom
3- storage for general affairs
4- sport equipment room
5- health room
6- duty room
7- canteen
8- bridge over artificial marsh
9- computer classroom
10- computer assistent room
11- music classroom
12- instrument room
13- painting room
14- reading room
15- multifunctional hall
16- nature classroom
17- nature preparation room
18- administration office

ASSEMBLED-STANDARD CONSTRUCTION -4
HOPE PRIMARY SCHOOL / BARKAM / CHINA

ECO-EDUCATION

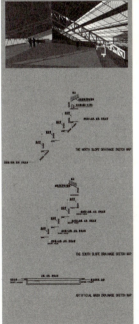

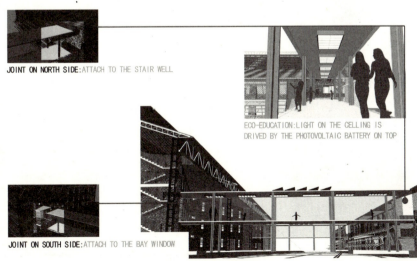

JOINT ON NORTH SIDE: ATTACH TO THE STAIR WELL

ECO-EDUCATION: LIGHT ON THE CEILING IS DRIVED BY THE PHOTOVOLTAIC BATTERY ON TOP

JOINT ON SOUTH SIDE: ATTACH TO THE BAY WINDOW

主要经济技术指标：	
总用地面积：	1.67公顷
总建筑面积：	7952平米
教学及教学辅助用房建筑面积：	3232平米
行政办公用房建筑面积：	345平米
生活服务用房建筑面积：	4375平米
道路广场面积：	4562平米
运动场用地：	3612平米
绿地面积：	2784平米
容积率：	0.48
绿地率：	21.3%
建筑密度：	34.3%
汽车泊位数：	6辆
自行车泊位数：	400辆

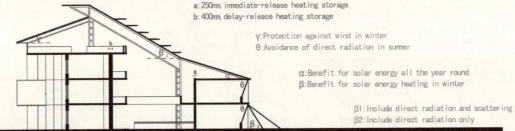

a: 250mm, immediate-release heating storage
b: 400mm, delay-release heating storage

γ: Protection against wind in winter
θ: Avoidance of direct radiation in summer

α: Benefit for solar energy all the year round
β: Benefit for solar energy heating in winter

β1: Include direct radiation and scattering
β2: Include direct radiation only

ASSEMBLED-STANDARD CONSTRUCTION -5
HOPE PRIMARY SCHOOL / BARKAM / CHINA

设计说明：

标准化 每个单元的剖面都是基于马尔康冬夏利用太阳能的最优化进行设计。同时，各装配部件都是标准化生产以适应震后快速建造。

适用性 平面根据功能和气候进行可适性设计，在不同情况下高效利用并节省能源。尽管剖面是标准化的，但平面在一定控制范围内仍然具有可变性和适用性。

可装配 每个单元都在设计中考虑了与其他部件连接的可能性，从而提供无限种装配的可能性以适应各种总平地形和功能。

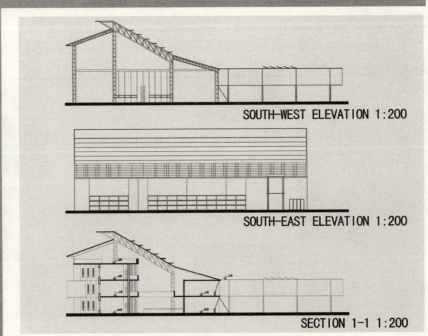

SOUTH-WEST ELEVATION 1:200
SOUTH-EAST ELEVATION 1:200
SECTION 1-1 1:200

SUMMER, IVY, SUNSHINE, WE LAUGH AND WE PLAY

IN WINTER, THE STREAT BECOMES SUNSHINE ATRIUM

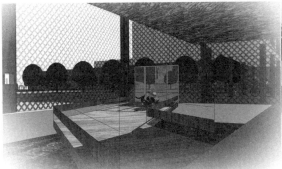
THE BRIDGE OVER ARTIFICIAL MARSH IS THE PLACE WE LOVE TO STAY

FROM PLAYGROUND WE CAN SEE OUR BEAUTIFUL CAMPUS

ASSEMBLED-STANDARD CONSTRUCTION -6
HOPE PRIMARY SCHOOL / BARKAM / CHINA

DESIGN INFORMATION:

STANDARDIZATION The section of each unit is designed under the consideration of making the best use of the solar energy at Barkam both in summer and winter. Meanwhile, the components are all standard-made to fit the fast pro-earthquake construction.

FLEXIBILITY The plan is created flexible with the thoughts of function and climates, which leads to highly effective utilization and saving ratio of energy on different occasions. Although the section is standard, the plan can be changed under a certain control.

ASSEMBLY Each unit is constructed with the possibility to connect with others, which provides infinite possibilities to assemble units in order to fit the location and function well.

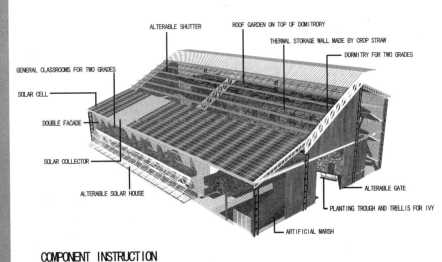

COMPONENT INSTRUCTION

技术专项奖
Prize for Technical Excellence

项目名称：农村阳光小学
Rural Sunshine Primary School

作　者：Osamu Morishita、Hidenori Tsuboi、Filippo Bari、Shingo Asazu、Satoko Furutani

参赛单位：Osamu Morishita, Architect & Associates

专家点评：

设计充分考虑了当地的自然条件，绿地和水塘的利用很好的维持了场地的现状，使学校周边保持自然生态的环境。缺点是被动设计概念不清晰，学生宿舍设计不利于管理。

Local natural conditions have been fully considered. The utilization of green field and pond makes the existing situation of the ground and natural ecological environment around the school kept very well. The shortcoming is that the concept of passive solar energy design is not clear and the dormitory of students is not good for management.

2009 Delta Cup
-International Solar Building Design Competition-

Rural Sunshine Primary School (in mian yang area)

A brief of the design scheme

Surrounded by rice fields and a natural pond, the teaching buildings are located in order to ensure the execution of school activities in a calm and natural environment.
The approach to the classroom is through ramps which connect the ground level to the teaching area.
Suspended passages link the common classrooms to special activities classrooms. The administrative area, at the boundaries of the sport field, is easily accessible from everywhere, and has the full visibility over all the school.
At the ground floor, the canteen is easily reachable both from the teaching area, both from the dormitories, which are located along the main street as a connection to the existing town.

The whole project fully makes use of all the natural resources available in site. The preexisting canals feed the pond and the rice field, which maintain the characterizing rural landscape and help having pleasant comfort conditions in summer. The preservation of trees, at the southern part of the area, contributes to not destroy natural conditions.
Building materials, mainly constituted by wood and concrete, are easy to get in the region. The construction methods and the absence of elaborated facilities, make the realization of the buildings simple and less expensive.

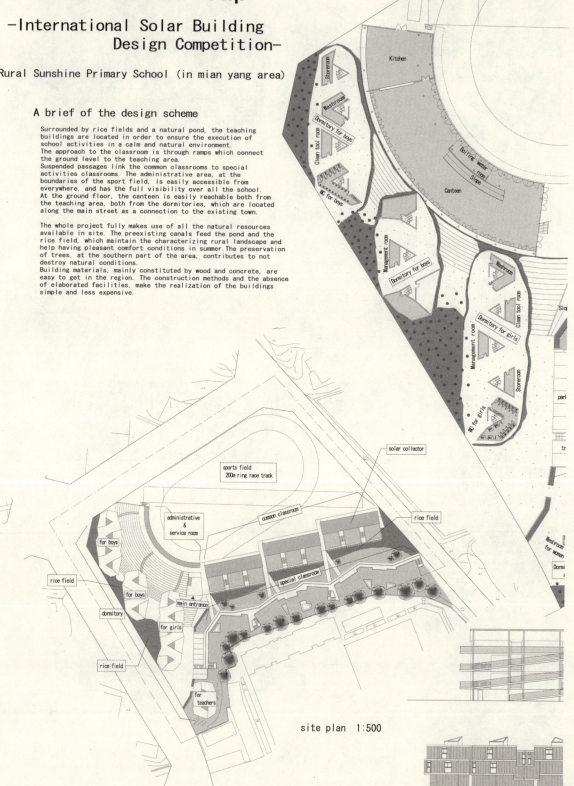

site plan 1:500

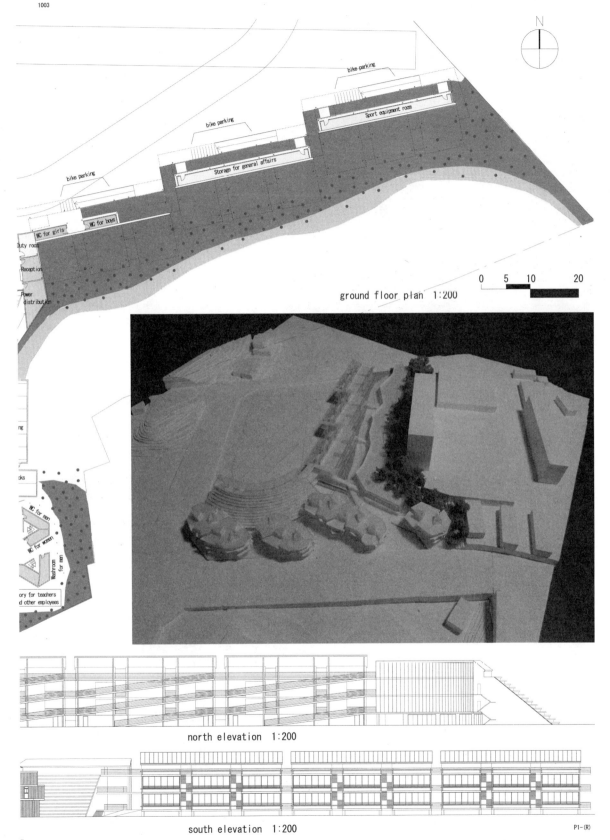

ground floor plan 1:200

north elevation 1:200

south elevation 1:200

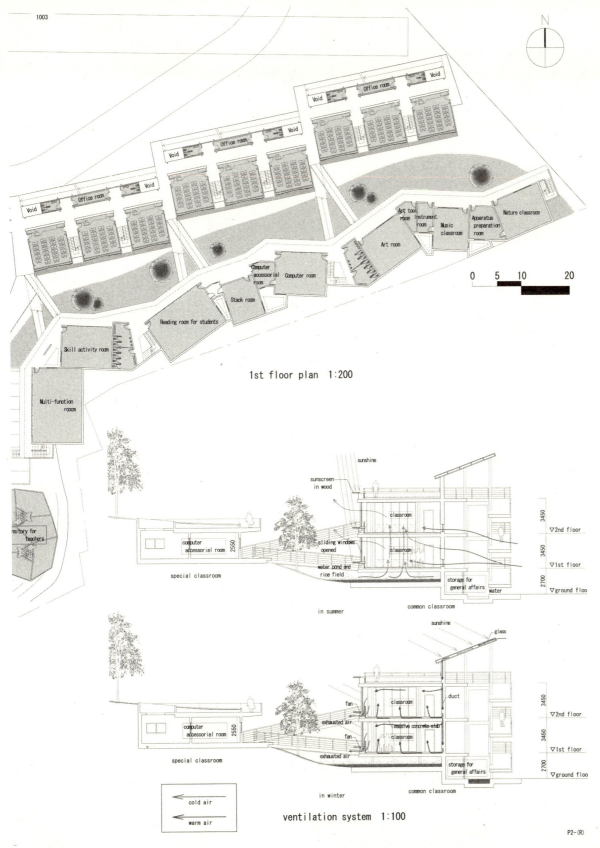

2009 Delta Cup
-International Solar Building Design Competition-

Serial number	item	unit	quantity
1	Total site area	hm²	1.67
2	Total floor area of the building	m²	7383
	Floor area of teaching and accessory rooms	m²	2016
	Floor area of administration	m²	391
	Floor area of service	m²	3170
3	Road & square area	m²	1971
4	Sport field	m²	5430
5	Greening area	m²	1522
6	Floor area ratio (excluding sport field)	%	0.81
7	Greening rate (excluding sport field)	%	0.11
8	Building density	%	0.21
9	Car parking spaces	car	6
10	Bike parking spaces	bike	78

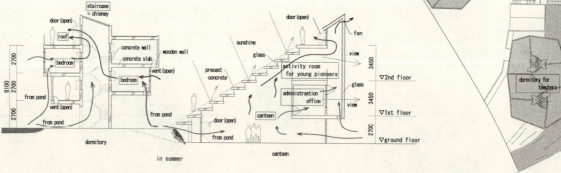

in summer

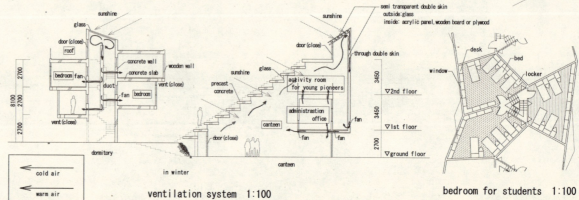

in winter

ventilation system 1:100

bedroom for students 1:100

2009台达杯国际太阳能建筑设计竞赛获奖作品集

2nd floor plan 1:200

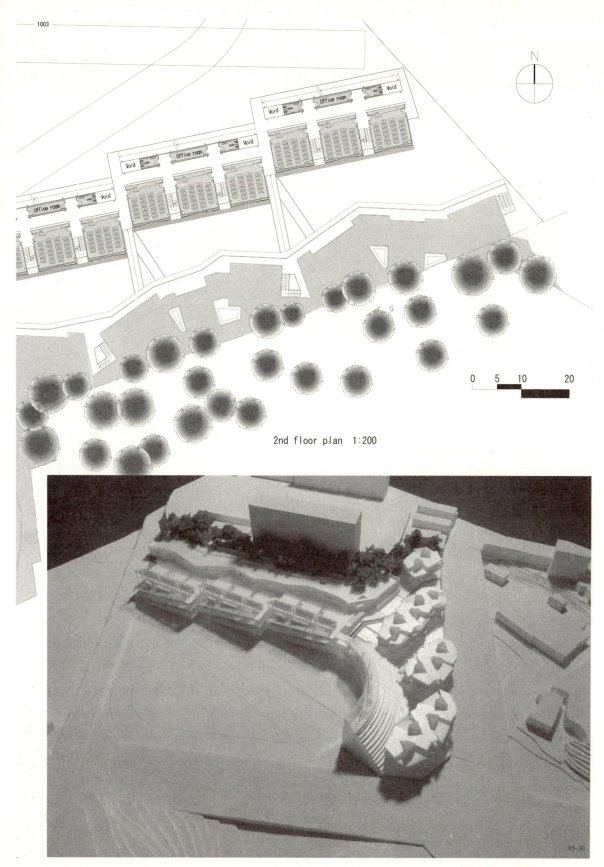

技术专项奖
Prize for Technical Excellence

项目名称：呼吸
　　　　　　Breathe
作　　者：申学峰、张仁子、金伟丽
参赛单位：吉林省延边大学科技技术学院

专家点评：
能量系统概念清晰，场地风环境分析到位，层层退台式的建筑形式可以很好地利用太阳能。缺点是建筑平面体量较大，形式单一。

The concept of energy system is clear and the analysis of wind environment of the ground is good. The building formed with step-type is profitable for solar utilization. The shortcoming is that the plan is too much expanded and the building form is monotonous.

BREATHE 1
breathe 1

设计过程：
因为绵阳地区湿度较高。
利用夏季通风来降低湿度，
且阻挡冬季通风来避免湿冷空气的侵袭。

Design process:
It shows high percentatge of relative humidity in MianYang city.
Make relative humidity lower in summer by ventilation, and try not to be affected by winter wind.

设计理念：
孩子们是我们的希望和未来，而学校则是给孩子们以梦想和希望的地方。
我们的设计理念是可以呼吸的学校。
无论动物还是植物，凡是活着的都在呼吸。
而呼吸则意味着它仍有希望追求它的梦想。
但问题是：建筑物会呼吸吗？
我们觉得：可以！

Main concept:
Children are our future and hope, and school is a place that gives children dreams and hope.
Our main concept is "The breathing school".
No matter what kind of things that is alive, such as animals or plants, they are all breathing.
Breath means he or she still has hope to dream his future.
The question is : Could a building breathe?
Our answer is : Yes!

夏季平均风速1.3m/s
Summer mean air velocity
冬季平均风速0.8m/s
Winter mean air velocity

夏季风通过建筑物
Summer wind gets through the building

用建筑物阻挡冬季风
Blocking the cold winter wind with this direction of building

以太阳辐射量最多的方向为温室的方向
Make greenspace to the direction of the most solar irradiation

利用烟囱效应，使温室里凉爽且氧气含量较高的空气通过教室流入到庭院。
Sucking cooled, much of oxygenous air from the greenspace through classrooms by chimney effect.

MAIN TECHNICAL AND ECONOMIC INDICATORS

Item	Quantity
Total site area	1.67 hm²
Total floor area of the building	10396.4 m²
Floor area of teaching and accessory rooms	3936.64 m²
Floor area of administration	554.34 m²
Floor area of service	5905.49 m²
Road & square area	7823.94 m²
Sport field	3616.06 m²
Greening area	1641 m²
Floor area ratio(excluding sport field)	79.46%
Greening rate(excluding sport field)	45.34%
Building density	21.67%
Car parking spaces	7 cars
Bike parking spaces	31 bikes

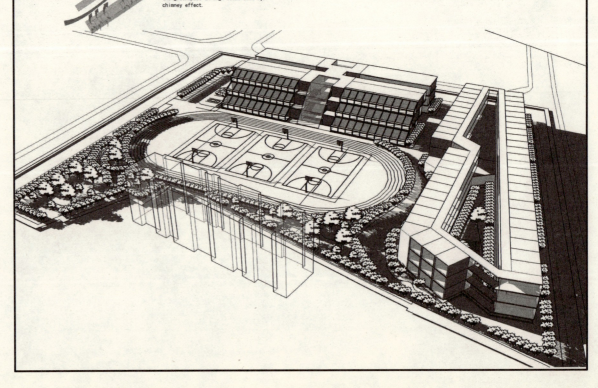

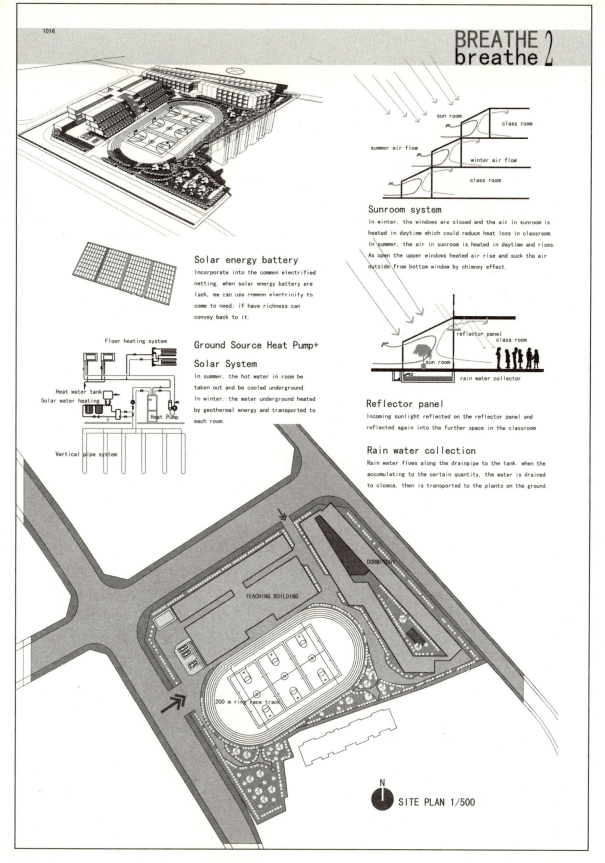

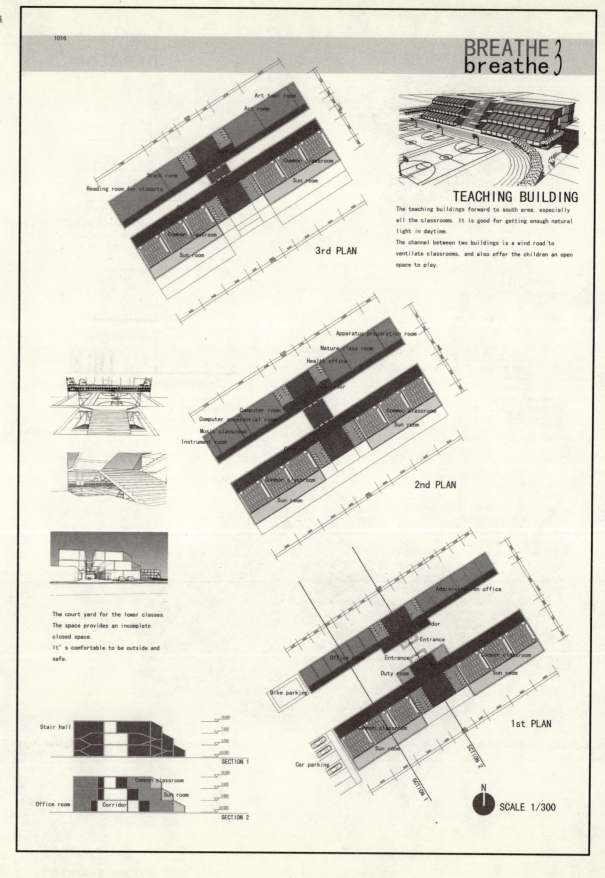

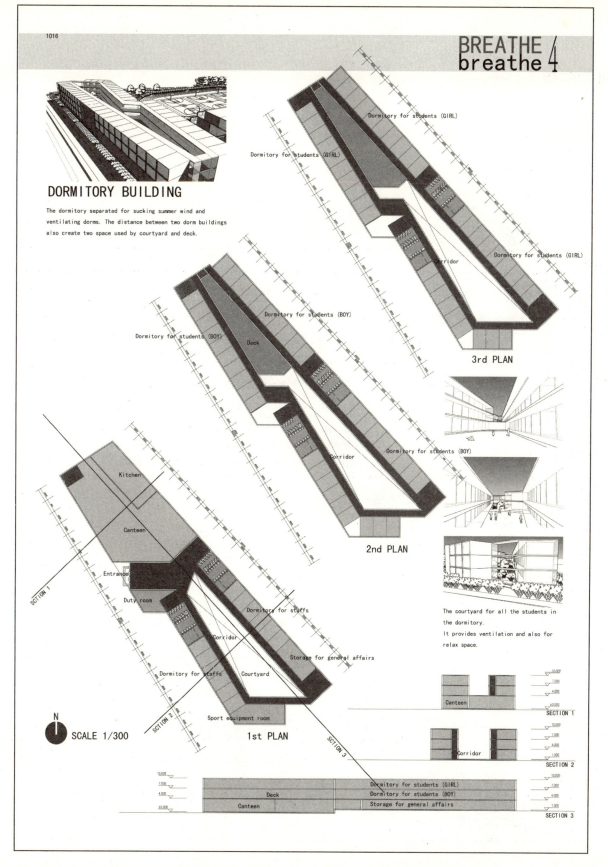

参赛人员名单
Name List of Attending Competitors

注册号	单位名称	作者
43	中国航空工业规划设计研究院	刘昌励
60	Jacobs	Jacobs
104	广州大学建筑与城市规划学院，广州市设计院	刘祖国、梅琬菲、文雅
108	重庆大学建筑城规学院	戈珍平、武子栋、陈东亮、刘晋
115	Périphériques Architectes、Paris FRANCE Ecole nationale supérieure architecture de Paris-La villette Paris FRANCE	Xiang WANG、Wenmu TIAN
127	西安欧亚学院艺术设计学院	陈涛、徐丽、马艺轩、李博超、杨波、岳磊
132	Popular Architecture LLC	Casey Mack、Konstantinos Stefanos Alivizatos、Artemis Papadatou、Tian Wei Jia、Yunfei Yang、Alice Chung、Jinchao Yuan
185	苏州市建筑设计研究院	朱涛
203	MV-ARCHITECTS码维建筑	唐勇
218	广州大学建筑与城市规划学院	陈粤、谢宝炫
249	西安建筑科技大学建筑学院	翟亮亮
255	内蒙古工业大学建筑学院、内蒙古新雅设计院	尚大为、马卿、杨朝辉、婷婷、赵晓娜
281	河北理工大学建筑工程学院	李子晗
287	Bureau d'achitecture A. Tecklenburg	TECKLENBURG André、TECKLENBURG Marius
297	青岛理工大学琴岛学院	张克青、王永晓
333	Bauhaus Dessau Foundation（德国德绍包豪斯研究院）	朱健、欧克男
335	西安建筑科技大学	高婉斐、刘海、王鹏
342	上海普塞建筑设计顾问有限公司	赵永晖、王建、王晓雯、张子良、黄玲杰、张蕾、李军平、赵嘉颖
344	Messana O'Rorke Architects	Brian Messana、Toby O'Rorke、Sungpyo Kim、Christopher Biggin
345	Gádor Luque. Sara Rojo. Víctor Quirós	Gádor Luque. Sara Rojo. Víctor Quirós
356	ATKINS 上海	戴一轩
362	重庆大学建筑城规学院	杨鹏程,黄一滔,李荣,吕家悦,何兆熊,谢崇实,冷泠
368	广东工业大学艺术设计学院	李土汉、陈锦添

注册号	单位名称	作者
370	西南交通大学建筑学院	汪海涛、张妍、姚亦梅、谭祈燕
371	燕山大学	雷炜、张泽
393	koopX	Joost Johannsen、Yun Yang、Fei Liu、Jianru Chen、Zhiyang Du
407	北京建筑工程学院	赵增鑫
414	Faculty of Architecture, Slovak University of Technology	Vančo Marek, Lucia Straková
420	淡江大学建筑系	张惇涵
422	河北工业大学	张超、侯薇、刘伟、朱江涛、孟庆山、唐一萌、李莺
433	上海大学美术学院建筑系	吴昊、姜成晟、王超、连晓俊、黄荣
436	重庆大学建筑城规学院	万博、刘菁
438	淡江大学建筑系	吴宛倩、王芊文
439	淡江大学建筑系	陈英南
451	滨州市规划设计研究院	张蕊、邓静静、张立杰
456	交通大学建筑研究所	曾哲文
457	台湾 交通大学 建筑研究所	吴忠勋
459	北京市建筑设计研究院	王鹏、于瑶
467	中国矿业大学南湖校区建工学院建筑系	林涛、刘茜
479	台湾交通大学建筑研究所	汪柏成
482	西安交通大学人居学院	班婧、张复昊、李江、胡涛
485	天津大学建筑学院、清华大学建筑学院	刘婷婷、刘烨、王鑫
497	politecnico di torino	simone tessa、Silvio Basso、Carlo Ostorero
505	青岛理工大学	李媛愿、黄靓、毕胜
513	同济大学建筑与城市规划学院	陈宇、吴维聪、王翔、苏岩芃、程冠华
524	华中科技大学建规学院	杜庆、赵森辉、毛一凡
526	华中科技大学建筑与城市规划学院绿色建筑研究中心	屈天鸣、贾子健、张仕寅
528	重庆大学建筑城规学院	郑文崇、谭志臣、林世华、李俊

注册号	单位名称	作者
534	华中科技大学建筑与城市规划学院	祝卿、聂子昊、刘旭明、黄晶、刘晖
541	内蒙古建筑职业技术学院、呼和浩特市勘察设计院	何牧、何晨旭
553	Architect n° 3230 - Ordine degli Architetti di Bologna (IT)	Nicola Bettini、Clara Masotti、Stefano Brunelli、Antonio Bandini、Stefano Massa、Debora Venturi
555	中联设计顾问有限公司	龚浩朋、龚浩斌
559	东南大学建筑学院	张饔洁、袁野
560	天津大学建筑学院	王剑威、田晓媛、王小荣
568	西安交通大学人居学院建筑系	王欣、杨柳
570	河北工业大学	李莺、张超、孟庆山、朱江涛、唐一萌
580	段晓明	段晓明
583	山东建筑大学	赵元博、杨济舟、王茹、刘聪
595	重庆大学建筑城规学院	林燕秋、陈晓宁
597	东南大学建筑学院	叶佳明、梁博、杨晓杉
598	重庆大学建筑城规学院	曾雪松、段希莹
600	西北工业大学建筑系	苏原、邹杰、任娟、李瀛
604	西安交通大学人居学院建筑系	王欣、杨柳
610	台湾交通大学建筑研究所	黄圣轩
612	甘肃省科学院太阳能建筑设计所	刘孝敏、刘叶瑞、吴红梅、田凌、徐平、吕萍秋、高昉
614	[abad] architects、Politecnico di Milano University、Ceretti & Tanfani S.p.A.、Associate of Sisa Engineering	Alessandro Bianchi、Andrea Pirollo、Massimiliano Zigoi、Alessandro Rogora、Roberto Siligardi、Giorgia Cantoni、Simona Marella、Claudio Sisa
617	清华大学建筑学院	郭梦笛、张梦瑶
618	清华大学建筑学院建筑学系	王朗、钟铮
621	sb architects	JOANA GOMES、GONÇALO CARVALHO、FRANCISCO SANTOS
623	Universidad del Valle de México	Federico Alcocer、Raul Uribe、Manuel Villarruel
625	Universidad del Valle de México	Ricardo Vazquez、Gustavo Martínez、Manuel Villarruel

注册号	单位名称	作者
626	清华大学建筑学院	韩天辞、张杨、崔乾旭
627	淡江大学	黄平玮、蔡静缇
629	西安交通大学	王永胜
630	华中科技大学建筑与城市规划学院	顾芳、刘碧峤
633	中国联合工程公司	姜传锹、孙玮、杜隽
635	东南大学建筑学院	包藏新、钟光浒
638	西安建筑科技大学	罗智星、谢栋、安赟刚、李琎、宋利伟、柯铠、陈晨
640	东南大学建筑学院	俞英、王欣、孟成、赵艳
657	河北燕大工程设计有限公司 燕园·天泽设计创作室、燕山大学	李凌高、孙喜山、李春雨、陈宁、车玉萍、包晶磊、纪鹏磊、冯志明
670	Karl-Johan Sellberg	Karl-Johan Sellberg
671	深圳大学建筑系	李鹏飞、陈聪、王睿、尚祖光
672	重庆大学建筑城规学院	祁钟、梁宇舒、董斌、黄河、张琦
677	清华大学建筑学院	夏君天、张愉
680	东南大学建筑学院	李鹏、郑彬、孙雪梅
687	深圳大学建筑与城市规划学院	关雪峰、蓝芬、封晨
688	东南大学建筑学院	王天晖,徐斌,丁瑜
694	重庆大学b区建筑城规学院	陈士群、梁海龙、王惠
700	Rafal Kolodziej、Edyta Nitecka、Liu Jian	Rafal Kolodziej architect
718	Nadir Bonaccorso Arquitectos Associados [nbAA]、Naturalworks、AdF engenheirose、consul tores	Nadir Bonaccorso、Vanda Alves、Joana Nunes、Clara Pereira、Gonçalo Macedo、João Barroso、Claudia Oliveira、Guilherme Carrilho da Graça、Cristina Horta、António Adão da Fonseca
722	东南大学建筑学院	丁瑜、徐斌、崔陇鹏、连小鑫
724	清华大学建筑学院	吴一凡、张婷、李倩怡、龚晨曦
729	东南大学建筑学院	王莎莎、高漫、刘泉泉、杨维菊、路可人
730	重庆大学建筑城规学院	何媛、施洁莹

注册号	单位名称	作者
733	重庆大学建筑城规学院	陈维果、王力
734	东南大学、南京工业大学	孙晓娟、谢礼祥
737	重庆大学建筑城规学院、重庆大学城市建设与环境工程学院	李荣、吕家悦、谢崇实、陈潇、吕晓田、杨鹏程、黄一滔、何兆熊、冷泠、陈佐球
742	东南大学建筑学院	杨宇、杨维菊、贾文娟
743	兰州理工大学设计艺术学院建筑系	庞达、黎江林、王天悦、葛曼
749	天津大学建筑学院	贾建、梁旭、陈筱
753	重庆大学建筑城规学院	李柏杨、朱韶华
762	中房集团建筑设计有限公司A+S设计机构、北京建筑工程学院建、北京筑都方圆建筑设计有限公司、汉森国际伯盛设计、北京越格建筑设计有限公司	史红昌、佟彬彬、田志伟、唐晓君、贾红星
764	石家庄铁道学院建筑与艺术分院、河北工业大学	高力强、邓可祥、刘瑞杰、欧阳文、黄帅、刘丹、丁磊、朱江涛
765	石家庄铁道学院建筑与艺术分院、河北工业大学	高力强、邓可祥、刘瑞杰、刘丹、丁磊、王兰、王帅、朱江涛
766	石家庄铁道学院建筑与艺术分院、河北工业大学	高力强、邓可祥、胡欣、王兰、王帅、欧阳文、黄帅、朱江涛
768	东南大学建筑学院	王冬、吴栋
777	University of British Columbia	Joe Yiu Ming Lee
780	重庆大学建筑城规学院	谷海东、王科
781	合肥工业大学建筑与艺术学院	熊鑫、胡朝昱、徐俊、刘璇、秦文
782	华中科技大学	万珊、宋迎
784	天津大学建筑学院、中建（北京）国际设计顾问有限公司	梁佳、张海涛
786	清华大学建筑学院	董超、程瑜
791	台湾交通大学建研所	林冠宇
792	沈阳建筑大学建筑与规划学院	任乃鑫、王磊、谢欢欢、毕岩、张轲、程广红
794	KOW International Dutch Design Consultants凯维建筑设计咨询（上海）有限公司	Tjerk Reijenga、蔡晓琦、张嫣、宋慧、郑欣、贺玮玮、杨丽、陈鸣
796	华南理工大学建筑学院	谢岱彬

注册号	单位名称	作者
798	IST	Ana Mestre、Ana Diogo、Lina Jesus、Joana Gonçalves
808	上海大学数码艺术学院	任飞、孙嘉威、霍昕、张蕙淼、章慧
810	东南大学建筑学院	刘铸、王强、张建伟
814	武汉大学城市设计学院建筑学系	易勇、刘子锷、区志勇、兰兵、鲍帆
816	北京市清华大学建筑系	刘瓅珣、胡若函
819	华中科技大学建筑与城市规划学院	李玲、余巍、申杰
822	东南大学建筑学院、南京大学建筑学院	周海龙、刘菲、吴昭华
824	同济大学	余中奇、赵梦桐
825	山东建筑大学建筑城规学院	张之光、康冬、薄超、宫月
832	中国恩菲工程技术有限公司	郭大力、王飞、李钊、王芳、曹亮
833	中国恩菲工程技术有限公司	郭大力、王博、李涛、张辛、侯舍辉、曹亮
837	沈阳都市建筑设计有限公司大连分公司	李芸、杨亚东、乔月环、曹李彬、衣佰文、唐大为
838	沈阳都市建筑设计有限公司大连分公司、大连民族学院	曹李彬、乔月环、李芸、衣佰文、赵春艳、刘潇轶、马琳、廖青、袁吉
839	沈阳建筑大学建筑与规划学院	任乃鑫、谢欢欢、王磊、张轲、程广红、毕岩
841	武汉理工大学土建学院	王华、杨艳丽
846	山东建筑大学建筑城规学院、天津大学建筑学院	郑瑾、袁天皎、李海波、郭聪
847	山东建筑大学建筑城规学院	焦尔桐、徐绍辉、张文超、黄砂、刘中爽、韩雪
848	山东建筑人学建筑城规学院	于娟、李燕、郭晨晨、陶然、宋京华、张玲
849	山东建筑大学建筑城规学院	司鸿斌、王卫超、王博成、李静
850	山东建筑大学建筑城规学院	刘雷超、高宁宁、李双凤、杨柳
851	山东建筑大学建筑城规学院	刘慧、孙瑞、贾培斌、李献良、张增武
853	山东建筑大学建筑城规学院	李善超、李金玲
854	山东建筑大学建筑城规学院	申建、周开济、相平、焦宇、李珂、刘亚朋
855	山东建筑大学建筑城规学院	曲羽、李楠、李崭、朱晓松
857	山东建筑大学建筑城规学院	孙铭、郑恒祥、杨磊、宋安、刘筱、杨琳

注册号	单位名称	作者
858	山东建筑大学	刘骞、张乐、吉喆、甄密、韦今吾、郑一林
859	山东建筑大学建筑城规学院	高逸嘉、王真、陈劭、曲艺、孙筱东、蒋会超、袁天骄
862	山东建筑大学建筑城规学院	韩超、许杰、梁叶、马邈、孟光、陈佳敏
863	山东建筑大学	李轶凡、王珣、李倩崔、鲁燕、陈冬梅、邓海剑
865	山东建筑大学	孙海洋、吴茱倩、张聪聪、牛艳丽、张国际、金昊宁、张磊
866	山东建筑大学建筑城规学院	房涛、刘震、李默、杨文江、吴昊、张乐岩
868	山东建筑大学建筑城规学院	李晓东、刘文、侯韵、孙燕怡、李媛媛、鞠晓磊、李浩田
870	山东建筑大学	黄雅文、李帅、田新、解婷、任艳
871	山东建筑大学建筑城规学院	李亮、刘杰明、马舒洁、王新彬、曹峰、康玉东、李明亮
879	Modasia	MAX FARAMOND、PAUL-ETIENNE GUILLERMIN、WEI TAO、ELISABETH GELIS
886	浙江大学建工学院建筑学系	郑斐，贾晶、燕艳、朱宇恒、葛坚
892	哈尔滨工业大学建筑学院	朱文桐
893	东南大学建筑学院、建筑设计研究院、无锡尚德	杨维菊、王俊豪、邵政、郑国、刘高鹏、傅秀章、刘俊、石邢、唐高亮、陈文华、李丽
894	重庆大学建筑城规学院	张振华、范臻
898	南京大学建筑学院、Waterloo University	沈开康、Jeffrey Cheng、张光伟
901	Tadamasa Kano Architectural Workshop、Takenaka Corporation、Hojo Structure Research Institute、Giken Engineer Network、Kobe University、Kyoto University of Foreign Studies、Osaka University of Arts	Adamasa Kano、Takeshi Matsuoka、Satoshi Noda、Koji Watanabe、Toshio Hojo、Muneaki Hashimoto、Mitsunori Wake、Sun Yuping、Peng Fei、J.Thomas Perry
911	大连民族学院	王玮龙、赵良雪、向声、左冰心
914	清华大学建筑学院	解扬、段文
920	哈尔滨工业大学建筑学院	赵巍
927	华中科技大学建筑与城市规划学院	石峰、李敏、彭鹏
930	江西师范大学城市建设学院	孙薇薇、涂健、梁家豪、陈星

注册号	单位名称	作者
934	江西师范大学城市建设学院	周飞、吕铭、辛淑萍、饶文超、吴铜亮、祝文凯
935	江西师范大学城市建设学院	邱小辉、但宣学、彭绍、鲍国强、方祺、洪晓棠
957	江西师范大学城市建设学院	杨云龙、王义辉、李利辉、闵明哲、高洪
959	内蒙古工业大学	赵晓娜、刘铮、韩超、马卿、白叶飞、聂广强、尚大伟
963	清华大学建筑学院	韩昊、贾崇俊
964	清华大学建筑学院	朱乃伟、贺储储
965	清华大学建筑学院	李岑、符传庆
972	河北工业大学建筑与艺术设计学院	孟庆山、唐一萌、张超、朱江涛、李莺、刘伟、朱赛鸿
973	河北工业大学	刘伟
974	河北工业大学建筑与艺术设计学院	唐一萌、孟庆山、朱江涛、刘伟、李莺、朱赛鸿
976	重庆大学建筑成规学院	李聪、杨聪
983	重庆大学建筑与城规学院	何兆熊、冷泠、杨鹏程、黄一滔、吕家悦、李荣、谢崇实
999	清华大学建筑学院	贾雪子、刘芸
1000	清华大学建筑系	刘平浩、王焓
1003	Osamu Morishita , Architect & Associates	Osamu Morishita、Hidenori Tsuboi、Filippo Bari、Shingo Asazu、Satoko Furutani
1008	西安科技大学建工学院	邱芃、梁钰、王垚、刘冬
1016	吉林省延边大学科技技术学院	申学峰、张仁子、金伟丽
1017	清华大学建筑系	陈茸、孙晨光、林波荣、王冰
1025	Islamic Azad University-Tabriz Branch	Ahadollah A'zami、Alireza Mehrfar、Asghar Motea Noparvari、Sanaz Mehrfar、Mazdak Shahed
1028	Islamic Azad University-Tabriz Branch	Ahadollah A'zami、Sadegh HasanNezhad、Raouf Jabbari
1031	Islamic Azad University-Tabriz Branch	Ahadollah A'zami、Aisan Baghvand、Maryam Bashirvandpolsangi、Baghvand
1032	Islamic Azad University-Tabriz Branch	Ahadollah A'zami、Reza Danandeh
1046	东南大学建筑学院	祝泮瑜、马溪茵
1047	台湾交通大学建筑研究所	叶千纶

注册号	单位名称	作者
1049	深圳建筑设计研究总院有限公司	李雷
1051	长安大学建筑学院	吴晓冬、王琼
1052	清华大学建筑学院建筑系	王飞、郑凯竞
1053	清华大学建筑学院	王申皓、于尧
1059	河南工业大学土木建筑学院	苏广
1076	浙江大学建筑工程学院建筑系	王竹、范理扬、朱怀、沈婷婷、陈宗炎
1077	清华大学建筑学院	倪小漪、王宇婧
1078	清华大学建筑学院建筑学系	王钰、刘伦

2009台达杯国际太阳能建筑设计竞赛办法
Competition Brief for International Solar Building Design Competition 2009

竞赛宗旨：

本次竞赛结合中国汶川地震的灾后重建工作，以"阳光与希望"为主题，向全球征集农村"阳光小学"设计方案，并把部分获奖方案在灾区付诸建设。

让我们以竞赛为平台，展示太阳能建筑技术，贯彻可持续运营理念，传递爱心与责任，播种梦想与希望。

竞赛题目：1.马尔康地区农村阳光小学；2.绵阳地区农村阳光小学

主办单位：国际太阳能学会
中国可再生能源学会

承办单位：国家住宅与居住环境工程技术研究中心
中国可再生能源学会太阳能建筑专业委员会

冠名单位：台达环境与教育基金会

评审专家：崔愷：国际建筑师协会副理事，中国建筑学会副理事长，中国国家工程设计大师，中国建筑设计研究院总建筑师。

Anne Grete Hestnes女士：前国际太阳能学会主席，挪威科技大学建筑系教授。

Deo Prasad：国际太阳能学会亚太区主席，澳大利亚新南威尔士大学建筑环境系教授。

Mitsuhiro Udagawa：国际太阳能学会日本区主席，日本早稻田大学博士，日本工学院大学建筑系教授。

GOAL OF COMPETITION

Combined with the reconstruction after Wenchuan earthquake in Sichuan Province, China, the competition with a theme of "Sunshine and Hope" is facing the whole world to collect design scheme of rural "Sunshine primary school" and some of awarded submissions will be put into construction in disaster area.

Taking the competition as a platform let's open out technology of solar building, promulgate the concept of sustainable operation, express our love and responsibility to the children in disaster area and make their life and studying filled with gladness, hope and genial sunshine.

THEMES OF COMPETITION:

1. Rural Sunshine Primary School in Ma Er Kang area;
2. Rural Sunshine Primary School in Mian Yang area.

ORGANIZER:

International Solar Energy Society
Chinese Renewable Energy Society

OPERATOR:

China National Engineering Research Center for Human Settlements
Special Committee of Solar Buildings, Chinese Renewable Energy Society

M.Norbert Fisch：德国不伦瑞克理工大学教授（TU Braunschweig），建筑与太阳能技术学院院长，德国斯图加特大学博士。

林宪德：台湾绿色建筑委员会主席，日本东京大学博士，台湾成功大学建筑系教授。

仲继寿：中国可再生能源学会太阳能建筑专业委员会主任委员，中国矿业大学博士。

喜文华：甘肃自然能源研究所所长，联合国工业发展组织国际太阳能技术促进转让中心主任，联合国可再生能源国际专家，国际协调员。

冯雅：中国建筑西南设计研究院副总工程师，中国建筑学会建筑热工与节能专业委员会副主任，重庆大学博士。

评比办法：

1. 由组委会审查参赛资格，并确定入围作品。
2. 由评委会评选出竞赛获奖作品。

评比标准：

1. 参赛作品须符合本竞赛"作品要求"的内容。
2. 鼓励创新，作品应体现原创性。
3. 设计作品应满足"附件3：农村阳光小学设计任务书"的要求，建筑技术与太阳能利用技术具有适配性。
4. 作品中应充分体现太阳能利用技术对降低建筑使用能耗的作用，并具有可实施性。
5. 作品应在经济可行、技术可靠的前提下，具有一定的超前性。
6. 评比指标解释：作品评分采用100分制。

评比指标	指标说明	分值
建筑设计	指规划布局、建筑构思、使用功能和建筑创新等方面	40
主动太阳能利用技术	通过专门设备收集、转换、传输、利用太阳能的技术，鼓励创新	10
被动太阳能利用技术	通过专门建筑设计与建筑构造利用太阳能的技术，鼓励创新	30
采用的其他技术	其他新能源利用技术和节水、节材、节地等方面技术，鼓励创新	10
技术的可操作性	技术的可行性、普及性和经济性要求	10

SPONSOR:

Delta Environmental & Educational Foundation

JURY MEMBERS:

Mr. Cui Kai, Deputy Board Member of IUA (International Union of Architects); Vice President of Architectural Society of China; National Design Master and Chief Architect of China Architecture Design & Research Group.

Ms. Anne Grete Hestnes, Former President of International Solar Energy Society and Professor of Department of Architecture, Norway Science & Technology University.

Mr. Deo Prasad, Asia-Pacific President of International Solar Energy Society (ISES) and Professor of Faculty of the Built Environment, University of New South Wales, Sydney, Australia.

Mr. Mitsuhiro Udagawa, President of ISES-Japan; Doctor of Engineering of Waseda University and professor of Department of Architecture, Kogakuin University.

Mr. M.Norbert Fisch, Professor of TU Braunschweig, President of the Institute of Architecture and Solar Energy Technology, Germany and Doctor of Stuttgart University, Germany.

Mr. Lin Xiande, President of Taiwan Green Building Committee; Doctor of Tokyo University, Japan and Professor of Faculty of Architecture of Success University, Taiwan.

Mr. Zhong Jishou, Chief Commissioner of Special Committee of Solar Building, Chinese Renewable Energy Society and Doctor of China University of Mining & Technology.

Mr. Xi Wenhua, Director-General of Gansu Natural Energy Research Institute; Director-General of UNIDO International Solar Energy Center for Technology Promotion and Transfer; expert in sustainable energy field from United Nations, international coordinator.

Mr. Feng Ya, deputy chief engineer of Southwest Architecture Design and Research Institute of China; deputy director of special committee of building thermal and energy efficiency, Architectural Society of China, Doctor of Chongqing University.

APPRAISAL METHODS:

1. Organizing Committee will check up eligible entries and confirm shortlist entries.
2. Jury will appraise and select out awarded works.

APPRAISAL STANDARD:

1. The entries must meet the demands of the Competition Requirement.
2. The entries should embody originality in order to encourage innovation.
3. The submission works should meet Annex 3 in Competition Brief. The building technology and solar energy technology should have adaptability to each other.
4. The submission works should play the role of reducing building energy

设计任务书及专业术语：（见附件）

1. 附件1：马尔康地区农村阳光小学气候条件
2. 附件2：绵阳地区农村阳光小学气候条件
3. 附件3：农村阳光小学设计任务书
4. 附件4：专业术语

奖项设置及奖励形式：

综合奖：获奖作品建筑设计与所选用太阳能技术具有较强的适配性。

一等奖作品 2名　　颁发奖杯、证书及人民币50000元奖金（税前）
二等奖作品 4名　　颁发奖杯、证书及人民币20000元奖金（税前）
三等奖作品 6名　　颁发奖杯、证书及人民币5000元奖金（税前）
优秀奖作品 40名　　颁发证书

技术专项奖：获奖作品在采用的技术或设计方面具有创新，实用性强。
获奖作品　名额不限　颁发证书

作品要求：

1. 在建筑设计方面应达到方案设计深度，在技术应用方面应有相关的技术图纸，作品图面、文字表达清楚，数据准确。
2. 作品基本内容包括：

（1）简要建筑方案设计说明（限200字以内），包括方案构思、太阳能综合应用技术与设计创新等。

（2）应按照附件3中所提供的"主要技术经济指标一览表"形式，编制相关技术经济指标。

（3）总平面图比例1：500（含场地及环境设计）。建筑设计内容为教学楼和宿舍楼，包括各层平面图、外立面图、剖面图，比例1：100～1：200（应能充分表达建筑与室内外环境的关系），重点部位、局部详图及节点大样比例自定，以及相关的技术图表等。

（4）建筑效果表现图1~3个。

（5）参赛者须将作品文件编排在840mm×590mm的展板区域内（统一采用竖向构图），作品张数应为2或4张。中英文统一使用黑体字。字体大小应符合下列要求：标题字高：25mm；一级标题字高：20mm；二级标题字高：15mm；图名字高：10mm；中文设计说明字高：8mm；英文设计说明字高：6mm；尺寸及标注字高：6mm。文件分辨率100dpi，格式为JPG或PDF文件。

3. 参赛者通过竞赛网页上传功能将作品递交竞赛组委会，入围作品由组委会统一编辑板眉、出图、制作展版。

4. 作品文字要求：除2.1"建筑方案设计说明"采用中英文外，其他为英文；尽量使用附件4中提供的专业术语。

consumption by utilization of solar energy technology and have feasibility.

5. The submission works should be advanced under the preconditions of economic practicability and technical liability.

6. A percentile score system is adopted for the appraisal.

APPRAISAL INDICATORS:

APPRAISAL INDICATOR	EXPLANATION	SCORES
Building design	Including layout planning, design ideas, usage function, architectural innovation and others.	40
Utilization of active solar energy technology	Technology concerning collecting, transforming, transmitting and utilizing solar energy by special equipments. Innovation is encouraged.	10
Utilization of passive solar energy technology	Technology of utilizing solar energy by special building design and construction. Innovation is encouraged.	30
Adoption of other technology	Other technology concerning new energy utilization, water saving, materials saving and land saving. Innovation is encouraged.	10
Operability of the technology	Feasibility, popularization of relevant technology and economy demands	10

THE TASK OF BUILDING DESIGN AND PROFESSIONAL GLOSSARY (Annex)

Annex 1: Climate conditions of the rural sunshine primary school in Ma Er Kang area
Annex 2: Climate conditions of the rural sunshine primary school in Mian Yang area
Annex 3: Task of building design of rural sunshine primary school
Annex 4: Professional Glossary

PRIZES:

General Prize :

Building design and selected solar energy technology must be excellent in adaptability to each other

First Prize: 2 winners.
The Trophy Cup, Certificate and Bonus RMB 50,000 (before tax) will be awarded.

Second Prize: 4 winners.
The Trophy Cup, Certificate and Bonus RMB 20,000 (before tax) will be awarded.

Third Prize: 6 winners.
The Trophy Cup, Certificate and Bonus RMB 5,000 (before tax) will be awarded.

Honorable Mention Prize: 40 winners.
The Certificate will be awarded.

参赛要求：

1. 欢迎建筑设计院、高等院校、研究机构、太阳能研发和生产企业等单位，组织建筑、结构、设备等专业的人员组成竞赛小组参加本次竞赛。

2. 请参赛人员访问www.isbdc.cn，按照规定步骤填写注册表，提交后会得到唯一的作品编号。一个作品对应一个注册号。提交作品时把注册号标注在每幅作品的左上角，字高6mm。注册时间2008年6月25日~2008年12月1日。

3. 参赛人员同意组委会公开刊登、出版、展览、应用其作品。

4. 被编入获奖作品集的作者，应配合组委会，按照出版要求对作品进行相应调整。

注意事项：

1. 参赛作品电子文档须在2009年1月18日前提交组委会，请参赛人员访问www.isbdc.cn，并上传文件，不接受其他递交方式。

2. 作品中不能出现任何与作者信息有关的标记内容，否则将视其为无效作品。

3. 组委会将及时在网上公布入选结果及评比情况，将获奖作品整理出版，并对获奖者予以表彰和奖励。

4. 获奖作品集首次出版后30日内，组委会向获奖作品的创作团队赠样书2册。

5. 竞赛消息公布，竞赛问题解答均可登陆竞赛网站查询。

所有权及版权声明：

1. 竞赛组委会享有刊登、出版、展览参赛作品的权利，并享有使用参赛作品于中国汶川地震灾区建设小学的权利。组委会在使用参赛作品时将对其作者予以署名，同时对作品将按出版或建设的要求做技术性处理。参赛作品均不退还。

2. 作者应对所提交作品的著作权承担责任，凡由于参赛作品而引发的著作权属纠纷均应由作者本人负责。

声明：

1. 参与本次竞赛的活动各方(包括参赛者、评委和组委)，即表明已接受上述要求。

2. 本次竞赛的参赛者，须接受评委会的评审决定作为最终竞赛结果。

3. 组委会对竞赛活动具有最终的解释权。

国际太阳能建筑设计竞赛组委会
网　　　址：www.isbdc.cn
组委会联系地址：北京市西城区车公庄大街19号
　　　　　　　国家住宅与居住环境工程技术研究中心，
　　　　　　　太阳能建筑专业委员会

PRIZE FOR TECHNICAL EXCELLENCE WORKS:

Prize works must be innovative with practicability in aspect of technology adopted or design.

The quota of it is open-ended. The Certificate will be awarded. In all cases the Jury's decision will be final.

REQIREMENTS OF THE WORK:

1. The work should reach the depth of scheme design level in building design and should be with relevant technical drawings in technology utilization. Drawings and text must be expressed clearly and its data must be mentioned exactly.

2. Basic contents of the work include:

(1) A brief of the design scheme (no more than 200 words) including scheme concept, general application technology of solar energy, design innovation, etc.

(2) Relevant technical and economic indicators expressed with a "list of technical and economic indicators" showed as a sample in annex 3.

(3) The general layout (1: 500, including site and environmental design), building plan (including teaching building and dormitory) of all floors, façade and section (1: 100~1: 200, which could fully show the building and its relationship, inner and outer), detail drawings for key parts and joints (scale is unlimited) and necessary figures or charts.

(4) Rendering perspective drawing (1-3).

(5) Participants should arrange the submission into two or four exhibition panels, each 840mm×590mm in size (arranged vertically). Font type should be in boldface. Font height is required as follows: title with 25mm; first subtitle with 20mm; second subtitle: 15mm; figure title: 10mm; design description in Chinese: 8mm, in English: 6mm; dimensions and labels: 6mm. File resolution: 100 dpi in JPEG or PDF format.

3. Participants should send (upload) a digital version of submission via FTP to the organizing committee, who will compile, print and make exhibition panels for all entries.

4. Text requirement: The brief of the design scheme (see 2.1) should be in Chinese and English, the others are in English. Participants should try to use the words from the Professional Glossary in Appendix 4.

PARTICIPATION REQUIREMENTS:

1. Institutes of architectural design, colleges and universities, research institutions and research and manufacture enterprises of solar energy are welcome to make competition groups with professionals of architecture, structure and equipment to attend the competition.

2. Please visit www.isbdc.cn. You may fill the registration list according to the instruction and gain an exclusive number of your work after submitting the list. One work only has one registration number. The number should be indicated in the top left corner of each submission work with height in 6mm. Registration time: 25th June, 2008 – 1st December, 2008.

3. Participants must agree that the Organizing Committee may publish, print, exhibit and apply their works in public.

4. The authors whose works are edited into the publication should cooperate with the Organizing Committee to adjust their works according to the requirements of printing.

邮　　　编：100044
联　系　人：郑晶茹、王　岩
联系电话：86-10-88327097，86-10-88327099
传　　　真：86-10-68302808
E-MAIL：info@isbdc.cn 、 info@house-china.net

附件1：

马尔康地区农村阳光小学气候条件

1．气象参数

北纬31.5°、东经102.2°、测量点海拔高度2666m

月份	空气温度 ℃	相对湿度 %	每日的太阳辐射量水平面 kW·h/m²/d	每日的太阳辐射量30°倾斜表面 kW·h/m²/d	大气压力 kPa	风速 m/s	土地温度 ℃
一月	-0.5	38.7	3.15	4.53	67.2	0.8	-4.6
二月	2.9	39.4	3.71	4.62	67.2	1.1	-2.6
三月	6.5	46.9	4.06	4.48	67.2	1.3	1.1
四月	9.8	53.6	4.77	4.78	67.4	1.4	5.3
五月	12.8	62.3	4.94	4.63	67.6	1.3	8.5
六月	15.1	72.8	4.60	4.21	67.6	1.0	11.4
七月	16.2	75.2	4.56	4.22	67.7	1.0	13.1
八月	15.6	74.7	4.36	4.23	67.8	0.9	12.4
九月	13.1	76.9	3.89	4.08	67.9	0.8	9.2
十月	9.3	71.3	3.52	4.16	67.9	0.8	5.0
十一月	3.9	52.7	3.27	4.52	67.7	0.7	0.0
十二月	-0.5	42.8	3.11	4.72	67.5	0.6	-3.4
年平均数	8.7	59.0	4.00	4.43	67.6	1.0	4.7

2．空调室外空气计算参数

	夏季	冬季
空气调节计算干球温度（℃）	27.3	-5.9
空气调节计算湿球温度（℃）：	17.3	——
空气调节计算日均温度（℃）	19.2	——
通风计算干球温度（℃）	22.5	-2.3
空气调节计算相对湿度（%）	51	39
平均风速（m/s）	1.2	1.0
风向	WNW	WNW

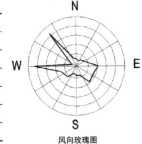

风向玫瑰图

ADDITIONAL ITEMS:

1. Participant's digital file must be uploaded to the organizing committee's FTP site (www.isbdc.cn) before 18th January, 2009. Other ways will not be accepted.

2. Any mark, sign or name related to participant's identity should not appear in, on or included with submission files, otherwise the submission will be deemed invalid.

3. The Organizing Committee will publicize the process and result of the appraisal online in a timely manner, compile and print publication of awarded works. The winners will be honored and awarded.

4. In 30 days after firstly published two of publication of award works will be freely presented by the Organizing Committee to the competition teams who are awarded.

5. The information concerning the competition as well as explanation about the competition may be checked and inquired in the website of the competition.

ANNOUNCEMENT ABOUT OWNERSHIP AND COPYRIGHT:

1. The Organizing Committee of the competition reserves the right of publishing, printing, and exhibiting the works, and also the right to use the submission works for school construction in Wenchuan earthquake area. When the works are used by the Organizing Committee, the names of authors will be affixed. In the mean while the works will be properly treated in technical according to the requirements of printing or project construction. All of submission works will be not returned back.

2. All authors must take responsibility for their copyrights of the works including all of disputes of copyright caused by the works.

ANNOUNCEMENT:

1. It implies that everybody who has attended the competition activities including participants, jury members and members of the Organizing Committee has accepted all requirements mentioned above.

2. All participants must accept the appraisal of the jury as the final result of the competition.

3. The Organizing Committee reserves final right to interpret for the competition activities.

Organizing Committee of International Solar Building Design Competition 2009

Website: www.isbdc.cn

Address of Organizing Committee:
China National Engineering Research Center for Human Settlements
Special Committee of Solar Buildings
No.19, Che Gong Zhuang Street,
Xi Cheng District,
Beijing, China
Post Code: 100044
Contact persons: Zheng Jingru, Wang Yan

3．冬季采暖室外空气计算参数

冬季采暖室外计算温度-3.9℃，冬季室外最多风向平均风速2.8m/s，最多风向WNW。

附件2：

绵阳地区农村阳光小学气候条件

1．气象参数

北纬31.5°、东经104.7°、测量点海拔高度472m

月份	空气温度	相对湿度	每日的太阳辐射量(水平面)	每日的太阳辐射量(30°倾斜表面)	大气压力	风速	土地温度
	℃	%	kW·h/m²/d	kW·h/m²/d	kPa	m/s	℃
一月	5.6	77.9	2.46	3.27	86.8	4.0	-0.9
二月	7.8	75.9	2.64	3.05	86.6	4.2	1.2
三月	11.6	73.0	3.04	3.22	86.4	4.4	6.0
四月	17.2	71.8	3.82	3.79	86.3	4.6	12.4
五月	22.1	67.9	4.18	3.92	86.1	4.4	17.2
六月	24.3	75.5	3.94	3.63	85.9	4.2	19.9
七月	26.1	80.5	4.04	3.75	85.8	3.6	21.8
八月	25.6	80.8	3.78	3.66	86.0	3.4	20.8
九月	21.6	81.7	2.94	3.00	86.4	3.9	17.1
十月	17.3	79.8	2.42	2.66	86.8	4.0	11.8
十一月	12.2	78.2	2.38	3.01	87.0	4.0	6.6
十二月	6.8	78.4	2.32	3.19	87.0	4.0	1.0
年平均数	16.6	76.8	3.17	3.35	86.4	4.1	11.3

2．空调室外空气计算参数

	夏季	冬季
空气调节计算干球温度（℃）	32.8	0.8
空气调节计算湿球温度（℃）：	26.3	—
空气调节计算日均温度（℃）	28.5	—
通风计算干球温度（℃）	29.3	2.9
空气调节计算相对湿度（%）	65	82
平均风速（m/s）	1.3	0.8
风向	WNW	ENE

风向玫瑰图

3．冬季采暖室外空气计算参数

冬季采暖室外计算温度2.6℃，冬季室外最多风向平均风速2.5m/s，最多风向ENE。

Tel. 86-10-88327097，86-10-88327099
Fax: 86-10-68302808
E-MAIL: info@isbdc.cn
info@house-china.net

ANNEX1:

Climate conditions of the rural sunshine primary school in Ma Er Kang area

1．D Basic climate conditions

North latitude 31.5°, east longitude 102.2°, measured point has height above sea level 2666 meters

Month	Air temperature	Relative humidity	Daily solar irradiation (horizontal)	Daily solar irradiation (30° slope surface)	Barometric	Air velocity	Ground surface temperature
	℃	%	kW·h/m²/d	kW·h/m²/d	kPa	m/s	℃
January	-0.5	38.7	3.15	4.53	67.2	0.8	-4.6
February	2.9	39.4	3.71	4.62	67.2	1.1	-2.6
March	6.5	46.9	4.06	4.48	67.2	1.3	1.1
April	9.8	53.6	4.77	4.78	67.4	1.4	5.3
May	12.8	62.3	4.94	4.63	67.6	1.3	8.5
June.	15.1	72.8	4.60	4.21	67.6	1.0	11.4
July	16.2	75.2	4.56	4.22	67.7	1.0	13.1
August	15.6	74.7	4.36	4.23	67.8	0.9	12.4
September	13.1	76.9	3.89	4.08	67.9	0.8	9.2
October	9.3	71.3	3.52	4.16	67.9	0.8	5.0
November	3.9	52.7	3.27	4.52	67.7	0.7	0.0
December	-0.5	42.8	3.11	4.72	67.5	0.6	-3.4
Average	8.7	59.0	4.00	4.43	67.6	1.0	4.7

2．Climate parameters for HVAC system design

	Summer	Winter
Outdoor air conditioning design dry bulb temperature (℃)	27.3	-5.9
Outdoor air conditioning design web bulb temperature (℃)	17.3	—
Daily mean air temperature (℃)	19.2	—
Dry bulb temperature of ventilation calculation (℃)	22.5	-2.3
Relative humidity of air conditioning calculation (%)	51	39
Mean air velocity (m/s)	1.2	1.0
Wind direction	WNW	WNW

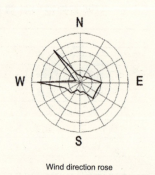

Wind direction rose

3．Climate parameters for heating system design in winter

附件3：

农村阳光小学设计任务书

农村阳光小学，规模18班，每班按40~45人计算，共计学生总数约为770人，住校学生约600人。学校用地形状见地形图（请下载附件5，AutoCAD格式文件），学校总占地面积为1.67hm²（含代征道路用地）；校舍由教学楼和配套用房组成，校舍建筑层数均不应超过3层，总建筑面积控制在8000m²以内,校舍使用面积分配见表一。设计要求如下：

1. 规划设计内容应包含学校用地范围内的全部区域。
2. 建筑设计内容为教学楼和宿舍楼，教学楼和学生宿舍宜采用框架结构。
3. 规划设计要求：
   (1) 学校规划应分区明确、功能合理、联系方便、互不干扰。
   (2) 教学用房和宿舍应满足冬至日底层满窗日照不小于2h的日照要求。
   (3) 处理好学生厕所与饮水位置，避免交通拥挤、气味外溢。
   (4) 学校体育场：
   跑道：200m环形跑道田径场；
   应设置适量的球类、器械等运动场（至少每6班设一个篮球场或排球场，足球场可根据条件设置）。
   (5) 集中设置绿地和室外自然科学园地，用地面积不宜小于1.5 m²/生，约1200 m²。
   (6) 自行车停放按0.1m²/在校生人数考虑，约77m²；汽车停车位按4辆小汽车、2辆中型货车考虑。
4. 教学楼建筑设计要求：
   (1) 教学楼包括教学及教学辅助用房和行政办公用房。
   (2) 普通教室及其他主要教学用房应有良好的自然采光和自然通风条件。
   (3) 教室应在满足使用功能的要求下，采用有效利用太阳能的平面形式。当采用矩形时，平面尺寸建议值（单位：mm）：矩形教室（轴线尺寸）：8400（长）×7200（宽）；布置方式见右侧示意图（单位mm）。

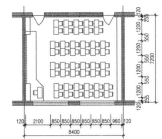

5. 宿舍楼建筑设计要求：
   (1) 宿舍楼包括学生和教工宿舍，不应与教学用房合建。
   (2) 学生宿舍由居室、管理室、盥洗室、厕所、贮藏室及清洁用具室组成，每间居室居住人数不宜多于8人。
   (3) 男女生宿舍应分区或分单元设置，出入口应分开设置。
   (4) 宿舍女厕按每12人一个大便器（或长1100mm的大便槽），男厕为20人一个大便器（或长1100mm的大便槽）和500mm长小便槽计算。厕所内应有洗手盆、污水池和地漏。
   (5) 教工宿舍每间居室按4人考虑。

Outdoor air conditioning design temperature: -3.9℃.
direction rose
Dominated wind direction WNW, with mean air velocity 2.8m/s

ANNEX 2：

Climate conditions of the rural sunshine primary school in Mian Yang area

1．Basic climate conditions

North latitude 31.5°, east longitude 104.7°, measured point has height above sea level 472 meters

Month	Air temperature	Relative humidity	Daily solar irradiation (horizontal)	Daily solar irradiation (30° slope surface)	Barometric	Air velocity	Ground surface temperature
	℃	%	kW·h/m²/d	kW·h/m²/d	kPa	m/s	℃
January	5.6	77.9	2.46	3.27	86.8	4.0	-0.9
February	7.8	75.9	2.64	3.05	86.6	4.2	1.2
March	11.6	73.0	3.04	3.22	86.4	4.4	6.0
April	17.2	71.8	3.82	3.79	86.3	4.6	12.4
May	22.1	67.9	4.18	3.92	86.1	4.4	17.2
June	24.3	75.5	3.94	3.63	85.9	4.2	19.9
July	26.1	80.5	4.04	3.75	85.8	3.6	21.8
August	25.6	80.8	3.78	3.66	86.0	3.4	20.8
September	21.6	81.7	2.94	3.00	86.4	3.9	17.1
October	17.3	79.8	2.42	2.66	86.8	4.0	11.8
November	12.2	78.2	2.38	3.01	87.0	4.0	6.6
December	6.8	78.4	2.32	3.19	87.0	4.0	1.0
Average	16.6	76.8	3.17	3.35	86.4	4.1	11.3

2．Climate parameters for HVAC system design

	Summer	Winter
Outdoor air conditioning design dry bulb temperature (℃)	32.8	0.8
Outdoor air conditioning design web bulb temperature (℃)	26.3	—
Daily mean air temperature (℃)	28.5	—
Dry bulb temperature of ventilation calculation (℃)	29.3	2.9
Relative humidity of air conditioning calculation (%)	65	82
Mean air velocity （m/s）	1.3	0.8
Wind direction	WNW	ENE

3．Climate parameters for heating system design in winter

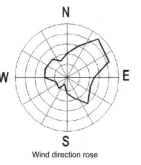

Wind direction rose

校舍使用面积参考指标（m²）

表1

序号及房间类别	用房名称	每间使用面积	间数	面积
教学及教学辅助用房				
1	*普通教室	56	18	1008
2	*音乐教室	56	1	56
3	*乐器室	18	1	18
4	*自然教室	80	1	80
5	*仪器准备室	40	1	40
6	*计算机教室	80	1	80
7	*计算机辅助室	18	1	18
8	多功能教室		1	150
9	*学生阅览室		1	120
10	*书库		1	52
11	劳技活动室		1	74
12	美术教室		1	80
13	美术教具室		1	18
14	*体育器材室		1	52
15	*教学办公用房	18	6	108
	小计：			1954
行政办公用房				
16	*行政办公用房	18	6	108
17	少先队活动室		1	52
18	*卫生保健室	18	1	18
19	*总务仓库		1	52
20	*值班室		1	18
	小计：			248
生活服务用房				
21	*食堂厨房		1	160
22	*食堂餐厅		1	500
23	*开水房			24
24	*学生宿舍			2000
25	*教工宿舍			300
26	*传达室		1	18
27	配电室		1	24
28	*厕所（教学区用）			170
	小计：			3196
	使用面积合计：			5398

注：*为必须设置内容

折算建筑面积：

建筑物名称	使用面积	K值（%）	建筑面积
1．教学及教学辅助用房	1954	65	3006
2．行政办公用房	248	65	382
3．生活服务用房	3196	70	4566
合计：	5398		7954

Outdoor air conditioning design temperature: 2.6℃.
Dominated wind direction ENE, with mean air velocity 2.5m/s

Wind direction rose

ANNEX 3:

Task of building design of rural sunshine primary school

An rural sunshine primary school has 18 classes and 40-45 students for each. Total students are around 770 and 600 of them are in residence. The shape of school site is showed in relief map (please download the AutoCAD file in Annex 5). Total area of the school is 1.67 ha. (including part of road outside the school). The school consists of teaching building, its accessory building and dormitory building and all of them are no more than three stories. Total floor area of the buildings should be restricted within 8000 m². The distribution of usable area of the school is showed in Table 1. The requirements of design are as follows.

☞ The planning and design should include the whole region of the school site.
☞ Building design of the school includes teaching building and dormitory building, which are suitable to adopt frame structure.
☞ The requirements of planning and design.
1. The planning and building design must be clear in division, reasonable in function and convenient to contact with each other but without disturbance to each other.
2. Teaching room and dormitory must meet the demand that the windows on first floor can get full sunlight for two hours on the day of the Winter Solstice (Dec.22nd).
3. The position of students' WC and drinking point should be arranged well in order to escape being crowded and bad smell diffusing out.
4. Sport field:
A track and sport ground with 200 m ring race track.
A certain area of field for balls and sport appliances should be arranged (at least every 6 classes have a court of basketball or volleyball and a football court can be set depending on the condition).
5. Green field and exterior court for natural science can be set concentratively. The area should be no less than 1.5 m² for one student, which is about 1200 m².
6. Bike parking space is set and calculated according to 0.1m² /student for all the students, which is 77m². Another parking space should be arranged for 4 cars and 2 trucks of mid size.

☞ Design requirements for teaching building
1. Teaching building includes teaching rooms, accessory rooms and administrative rooms.
2. Common classrooms and other main teaching rooms should have good condition of daylighting and natural ventilation.
3. Under the precondition of meeting the demands of use function, the classrooms should adopt such plan types which are effective to utilize solar energy. When rectangle room is adopted the

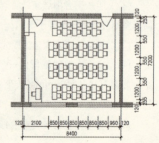

主要技术经济指标一览表（需要在作品中列出）　　　表2

序号	名称	单位	数量
1	总用地面积	hm²	1.67
2	总建筑面积	m²	
	教学及教学辅助用房建筑面积	m²	
	行政办公用房建筑面积	m²	
	生活服务用房建筑面积	m²	
3	道路广场面积	m²	
4	运动场用地	m²	
5	绿地面积	m²	
6	容积率（不含运动场用地）		
7	绿地率（不含运动场用地）	%	
8	建筑密度	%	
9	汽车泊位数	辆	
10	自行车停车数	辆	

附件4：

术　语

百叶通风	— shutter ventilation
保健室	— health room
保温	— thermal insulation
被动太阳能利用	— passive solar energy utilization
敞开系统	— open system
除湿系统	— dehumidification system
储热器	— thermal storage
储水量	— water storage capacity
穿堂风	— through-draught
传达室	— reception room
窗墙面积比	— area ratio of window to wall
次入口	— secondary entrance
单身教工宿舍	— hostel for single staffs
导热系数	— thermal conductivity
低能耗	— lower energy consumption
低温热水地板辐射供暖	— low temperature hot water floor radiant heating
地板辐射采暖	— floor panel heating
地面层	— ground layer
电化教室	— audio-visual room
额定工作压力	— nominal working pressure

dimensions of it are suggested as 8400×7200mm (axis to axis). Its arrangement is showed as the right side sketch drawing.

☞ Design requirements for dormitory building

1. Dormitory buildings are for students and teachers and other employees, which should not be constructed with teaching building together.
2. Student dormitory consists of bedrooms, management rooms, washrooms, WC, storeroom and clean tool room. Every bedroom is for eight persons in maximum.
3. Dormitories for boys and girls as well as their passageway should be arranged separately in different areas or different units.
4. In girls' WC, a toilet bowl (or 1100mm long water flume) is set for 12 persons. In boys' WC, a toilet bowl (or 1100mm long water flume) and 500mm long urinal flume are set for 20 persons. Also a washbowl, slop sink and drainer should be included in WC.
5. Dormitory for teachers and other employees should be designed as one room for four persons.

TABLE1 REFERENCE INDICATORS OF USABLE AREAS OF THE SCHOOL (m²)

SERIAL NUMBER	ITEM	USABLE AREA / ROOM	QUANTITY	TOTAL AREA
Teaching and accessory rooms				
1	* Common classroom	56	18	1008
2	* Music classroom	56	1	56
3	* Instrument room	18	1	18
4	* Nature class room	80	1	80
5	* Apparatus preparation room	40	1	40
6	* Computer room	80	1	80
7	* Computer accessorial room	18	1	18
8	Multi-function room		1	150
9	* Reading room for students		1	120
10	* Stack room		1	52
11	Skill activity room		1	74
12	Art room		1	80
13	Art tool room		1	18
14	* Sport equipment room		1	52
15	* Office room	18	6	108
	Subtotal			1954
Administrative room				
16	* Administration office	18	6	108
17	Activity room for Young Pioneers		1	52

防潮层	— wetproof layer
防冻	— freeze protection
防水层	— waterproof layer
分户热计量	— household-based heat metering
分离式系统	— remote storage system
风速分布	— wind speed distribution
封闭系统	— closed system
辅助热源	— auxiliary thermal source
辅助入口	— accessory entrance
高层住宅	— multi-storey dwelling
隔热层	— heat insulating layer
隔热窗户	— heat insulation window
跟踪集热器	— tracking collector
光伏发电系统	— photovoltaic system
光伏幕墙	— PV façade
广播室	— radio room
回流系统	— drainback system
回收年限	— payback time
集热器瞬时效率	— instantaneous collector efficiency
集热器阵列	— collector array
集中供暖	— central heating
间接系统	— indirect system
建筑节能率	— building energy saving rate
建筑密度	— building density
建筑面积	— building area
建筑物耗热量指标	— index of building heat loss
教工厕所	— toilet for teachers and other staffs
教工食堂	— canteen for teachers and other staffs
教师办公室	— teacher office
教师阅览室	— reading room for teachers
教学楼	— teaching building
教学用房	— teaching room
节能措施	— energy saving method
节能量	— quantity of energy saving
紧凑式太阳热水器	— close-coupled solar water heater
经济分析	— economic analysis
卷帘外遮阳系统	— roller shutter sun shading system
开水房	— boiling water room
科技活动室	— room for science and technical activities
空气集热器	— air collector

SERIAL NUMBER	ITEM	USABLE AREA / ROOM	QUANTITY	TOTAL AREA
18	* Health office	18	1	18
19	* Storage for general affairs		1	52
20	* Duty room		1	18
	Subtotal			248
Service room				
21	* Kitchen		1	160
22	* Canteen		1	500
23	* Boiling water room		1	24
24	* Dormitory for students			2000
25	* Dormitory for staffs			300
26	* Reception		1	18
27	Power distribution		1	24
28	* WC (for teaching area)			170
	Subtotal			3196
	Total usable area			5398

* must be established.

FLOOR AREA OF BUILDINGS CONVERTED

Type	Usable area	K (%)	Building area
Teaching and accessory building	1954	65	3006
Administrative building	248	65	382
Service building	3196	70	4566
Total	5398		7954

TABLE2 MAIN TECHNICAL AND ECONOMIC INDICATORS (Need to be shown in final submission)

Serial number	item	unit	quantity
1	Total site area	hm^2	1.67
2	Total floor area of the building	m^2	
	Floor area of teaching and accessory rooms	m^2	
	Floor area of administration	m^2	
	Floor area of service	m^2	
3	Road & square area	m^2	
4	Sport field	m^2	
5	Greening area	m^2	
6	Floor area ratio(excluding sport field)	%	
7	Greening rate(excluding sport field)	%	
8	Building density	%	
9	Car parking spaces	car	
10	Bike parking spaces	bike	

空气质量检测	– air quality test (AQT)
篮球场	– basketball field
劳动技术教室	– labor skill room
乐器室	– instrument room
立体绿化	– tridimensional virescence
绿地率	– greening rate
毛细管辐射	– capillary radiation
木工修理室	– repairing room for woodworker
耐用指标	– permanent index
能量储存和回收系统	– energy storage & heat recovery system
排球场	– volleyball field
平屋面	– plane roof
坡屋面	– sloping roof
普通教室	– common classroom
强制循环系统	– forced circulation system
热泵供暖	– heat pump heat supply
热量计量装置	– heat metering device
热稳定性	– thermal stability
热效率曲线	– thermal efficiency curve
热压	– thermal pressure
人工湿地效应	– artificial marsh effect
日照标准	– insolation standard
容积率	– floor area ratio
三联供	– triple co-generation
设计使用年限	– design working life
社团办公室	– association office
实验室	– laboratory
使用面积	– usable area
室内舒适度	– indoor comfort level
书库	– book storehouse
双层幕墙	– double façade building
太阳方位角	– solar azimuth
太阳房	– solar house
太阳辐射热	– solar radiant heat
太阳辐射热吸收系数	– absorptance for solar radiation
太阳高度角	– solar altitude
太阳能保证率	– solar fraction
太阳能带辅助热源系统	– solar plus supplementary system
太阳能电池	– solar cell
太阳能集热器	– solar collector
太阳能驱动吸附式制冷	– solar driven desiccant evaporative cooling
太阳能驱动吸收式制冷	– solar driven absorption cooling
太阳能热水器	– solar water heating
太阳能烟囱	– solar chimney
太阳能预热系统	– solar preheat system
太阳墙	– solar wall
体育器械室	– sport equipment room
田径场	– track and field ground
填充层	– fill up layer
通风模拟	– ventilation simulation
外窗隔热系统	– external windows insulation system
温差控制器	– differential temperature controller
屋顶植被	– roof planting
屋面隔热系统	– roof insulation system
相变材料	– phase change material (PCM)
相变太阳能系统	– phase change solar system
相变蓄热	– phase change thermal storage
行政办公室	– administration office
行政用房	– administrative room
蓄热特性	– thermal storage characteristic
学生厕所	– Toilet for students
学生阅览室	– reading room for students
音乐教室	– music classroom
雨水收集	– rain water collection
浴室	– bathroom
运动场地	– schoolyard
遮阳系数	– sunshading coefficient
直接系统	– direct system
值班室	– duty room
智能建筑控制系统	– building intelligent control system
中庭采光	– atrium lighting
主入口	– main entrance
贮热水箱	– heat storage tank
准备室	– preparation room
准稳态	– quasi-steady state
自然通风	– natural ventilation
自然循环系统	– natural circulation system
自行车棚	– bike parking
总务库	– storage for general affairs

空气质量检测	— air quality test (AQT)	太阳能驱动吸附式制冷	— solar driven desiccant evaporative cooling
篮球场	— basketball field	太阳能驱动吸收式制冷	— solar driven absorption cooling
劳动技术教室	— labor skill room	太阳能热水器	— solar water heating
乐器室	— instrument room	太阳能烟囱	— solar chimney
立体绿化	— tridimensional virescence	太阳能预热系统	— solar preheat system
绿地率	— greening rate	太阳墙	— solar wall
毛细管辐射	— capillary radiation	体育器械室	— sport equipment room
木工修理室	— repairing room for woodworker	田径场	— track and field ground
耐用指标	— permanent index	填充层	— fill up layer
能量储存和回收系统	— energy storage & heat recovery system	通风模拟	— ventilation simulation
		外窗隔热系统	— external windows insulation system
平屋面	— plane roof	温差控制器	— differential temperature controller
坡屋面	— sloping roof	屋顶植被	— roof planting
普通教室	— common classroom		
热量计量装置	— heat metering device	相变蓄热	— phase change thermal storage
热稳定性	— thermal stability	行政办公室	— administration office
热效率曲线	— thermal efficiency curve	行政用房	— administrative room
热压	— thermal pressure	蓄热特性	— thermal storage characteristic
人工湿地效应	— artificial marsh effect	学生厕所	— Toilet for students
日照标准	— insolation standard	学生阅览室	— reading room for students
容积率	— floor area ratio	音乐教室	— music classroom
三联供	— triple co-generation	雨水收集	— rain water collection
设计使用年限	— design working life	浴室	— bathroom
社团办公室	— association office	运动场地	— schoolyard
		遮阳系数	— sunshading coefficient
书库	— book storehouse	智能照明控制系统	— lighting intelligent control system
双层幕墙	— double facade building	中庭采光	— atrium lighting
太阳方位角	— solar azimuth	主入口	— main entrance
太阳房	— solar house	贮热水箱	— heat storage tank
太阳辐射热	— solar radiant heat	准备室	— preparation room
太阳辐射热吸收系数	— absorptance for solar radiation	准稳态	— quasi-steady state
太阳高度角	— solar altitude	自然通风	— natural ventilation
太阳能保证率	— solar fraction	自然循环系统	— natural circulation system
太阳能电池	— solar cell	总务库	— storage for general affairs
太阳能集热器	— solar collector		

图4 学校内院透视图
Figure 4 Interior of the school

图5 教室通风示意图
Figure 5 Classroom ventilation

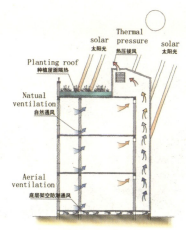
图6 宿舍太阳能烟筒通风示意图
Figure 6 Solar chimney for ventilation in hostel

校园绿化景观统一规划设计，合理利用原有的地形地貌，建筑与自然环境有机结合，营造绿色校园，并实现了无障碍设计。

学校在规划设计时，充分考虑了绵阳地区的气候特点和校园能源需求的特点，结合场地特点和不同建筑的使用功能，合理优化平面规划布局和建筑空间。充分利用太阳能等可再生能源，特别是采用太阳能被动利用技术，传播节能减排理念，营造自然生态环境。主要反映在以下几个方面。

一、被动式太阳能建筑设计

绵阳地区夏季闷热、冬季潮湿。结合教学楼、宿舍楼和食堂等建筑使用特点，采用了夏季以通风、隔热、遮阳、隔潮为主，冬季以集热、保温、通风为主的被动设计技术。主要技术策略如下：

1. 建筑围护结构的节能设计

采用规整的建筑体形和低成本围护结构保温隔热体系。建筑外围护墙体采用复合墙体，在围护结构外侧复合保温材料，提高墙体的保温隔热性能；屋面除设置复合保温材料外，还设置了空气缓冲层。

2. 被动式通风技术设计

1）校园规划及建筑群体风环境组织

根据绵阳地区的风速和主导风向，考虑周边地形地貌对校园风环境的影响，利用风环境模拟技术对校园规划和建筑群体的空间组合进行优化设计。建筑空间组合形式营造出有利于夏季自然通风、冬季阻挡冷风的校园风环境。

2）建筑单体通风设计

采用单廊式平面布局，充分利用自然通风。顶层教室结合屋面造型设置天窗，加强顶部房间通风，改善顶部房间夏季湿热环境。教学辅助办公区楼梯间及宿舍南侧设置强化的太阳能烟筒提高室内通风能力。

boiling water room and others).

In sport and rest area there are a playground with a 200 meter raceway, two of basketball court, ping-pong court and a natural science garden of ecological marsh type.

In the planning design of campus landscape, the original landform is fully utilized and buildings are well combined with natural environment, thus making a green campus. Besides non-obstacle design is realized as well.

In the planning and design of the school, the characteristics of the climate in Mianyang and energy needs of the campus are fully considered. Combined with the character of the site and functions of every building, the layout and arrangement of architectural spaces are rationally optimized. Solar energy and other renewable energies are fully utilized, especially on the adoption of passive solar technology which may diffusing the idea of saving energy, reducing the emissions of harmful gas and discharge of sewerage and making a natural ecological environment. They are showed in following aspects.

1 Building design with passive solar technology

In Mianyang it is muggy in the summer and wet in the winter. Combined with the functions of teaching building, hostel building and canteen, passive solar technology is adopted. Mainly it includes ventilation, isolation from hotness and moisture and sunshade in the summer while collecting heat energy, keeping warm and ventilation in the winter.

1) Design of energy saving for peripheral structure of the buildings

The buildings are designed with regular and simple form and a low cost isolation system is adopted for peripheral structure. Exterior walls are of compound ones with isolated materials in order to upgrade isolation performance of the walls. For roofs besides the adoption of isolation materials, an air cushioned layer is designed as well.

2) Passive ventilation technology design

☞ **Organization of wind environment for the campus planning and building groups**

Based on the speed and dominate direction of the wind in Mianyang,

图7 宿舍楼顶种植屋面示意图
Figure 7 Planting roof of the hostel

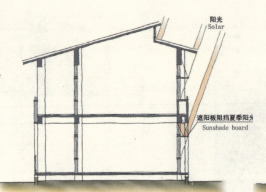

图8 利用遮阳板（反射板）阻挡夏季阳光
Figure 8 Shelter sun light in the winter

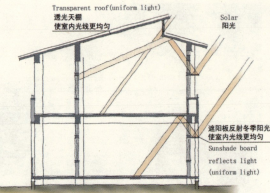

图9 利用反射板（遮阳板）调节室内光线
Figure 9 Adjust interior light

3. 被动式隔热设计

在学校的教室、办公和宿舍等主要建筑的屋顶部分设置热缓冲层，加强隔热和保温作用。在宿舍楼顶设置种植屋面，在保温隔热作用的同时，为学生营造出另一个趣味活动空间。

4. 建筑遮阳设计

教室南侧外窗的上部1/3处设置遮光板，结合屋檐形成完整的遮阳体系，充分利用绵阳地区夏、冬季日照高度角的差别，起到夏季遮阳、冬季透光的作用。

宿舍利用南侧设置阳台，除满足日常生活使用外，还具有建筑遮阳功能。教工宿舍的阳台在冬季还具有阳光间作用。

5. 教室光线调节设计

教室南侧外窗上部的反射板与遮阳板合并设置，能将透过上部窗户的光线通过反射板和教室天棚反射到教室的深处，营造更加均匀的室内光环境。

6. 隔潮设计

在首层房间地面下设置300mm的架空层，使之与地面形成空气流动层，有效防止地面湿气进入室内，达到隔潮作用。

considering the influence of the landform around the school to the wind environment of the campus, the planning and space composition of the buildings are optimized in design by simulation technology of wind environment. The space composition of buildings makes such a good wind environment profitable for natural ventilation in the summer and obstruction from cold wind in the winter.

☞ **Ventilation design for buildings**

Buildings are arranged with side corridors to fully utilize natural ventilation. For classrooms on the top floor some skylight are used to strengthen ventilation and improve their wet environment in the summer. In the stair case of teaching and working area and on the southern side of the hostel, strengthened solar chimneys are set up to increase the capacity of interior ventilation.

☞ **Passive isolation design**

In the roofs of main buildings as classrooms, offices and hostel heat cushioned layers are set up to strengthen the role of isolation and keeping warm. Planting roof is used in hostel building which may also make an interesting space for students besides its isolation role.

☞ **Sunshade design of the buildings**

On the upper part of the windows on southern side of the classrooms sunshades are set up which makes a whole sunshade system combined with the eaves and fully utilized the difference of the shining angles of the sun to play the role of sunshade in the summer and gaining sunshine in the winter.

Besides the use for daily life the balconies on the southern side of the hostel may play the role of sunshade. Also the balconies in the hostel for teachers and others can play the role as a sun room in the winter.

☞ **Light adjustment design of classrooms**

Light-reflection boards and sunshade boards are set jointly on upper part of the southern windows of the classrooms. The light entering from upper part of the window will be reflected to the deeper place of the room via light-reflection boards and ceilings, thus making interior light more uniform.

☞ **Design of isolation from moisture**

Under the ground of first floor an aerial space with the height of 300mm is set up for air flowing in order to avoid wet air on the ground outside entering the room and play the role of isolation from moisture.

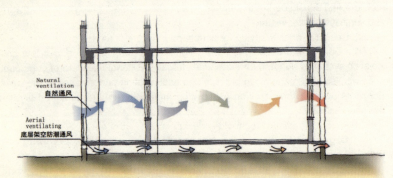

图10 利用架空层起到隔潮作用
Figure 10 Aerial space under the floor for isolation from moisture

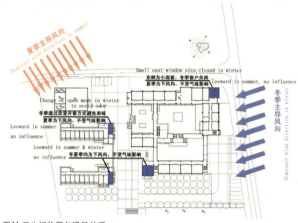

图11 卫生间位置与通风关系
Figure 11 Relation between the position of toilets and ventilation

图12 食堂屋顶太阳能热水及太阳能开水系统
Figure 12 Solar water heating and boiling water systems on the canteen roof

7. 地下室自然采光设计

在教学辅助办公区的半地下空间的西侧设置窗井，既能满足自然采光的要求，同时又可改善半地下室的通风问题，起到地下室除湿的作用。

8. 卫生间通风屏蔽设计

首先在卫生间位置的选择上，优先布置在冬夏两季主导风向的下风向。在不能满足两个季节都是下风向时，冬季和夏季采用不同开窗方式，避免串味。宿舍楼卫生间采用的特殊的开窗设计，阻挡冬季冷风直接吹进厕所内，并通过在窗外形成的涡流带走卫生间内的气味。

二、主动式太阳能建筑设计

考虑到绵阳地区太阳能辐照条件，在食堂设计集中学生浴室。采用太阳能集中热水系统为住宿学生和教工提供淋浴所需的生活热水。利用太阳能开水系统为师生提供开水。

三、生态湿地与生活污水处理

利用学校的地形特征，将生活污水无动力收集至校园西南生态湿地附近。采用人工湿地生态处理技术对生活污水进行处理，达到零排放。

☞ Natural lighting design for basement

In teaching and working area an areaway is arranged on the west side of semi-basement, which may meet the requirement of natural lighting and improve the ventilation of the basement as well in order to exhaust the moisture out.

☞ Shield design for toilet ventilation

Firstly the position of the toilets is chosen on the leeward of main wind direction in the summer and winter. In case they can not be arranged on the leeward in both seasons the utilization of different way to open the window will avoid bad odor flowing to other rooms. In the hostel a special design for opening windows will avoid cold wind entering in the winter and to make a whirlpool outside the window, thus bringing bad smell away.

2 Active solar technology in building design

Considering the condition of solar radiation in Mianyang, student bathrooms are arranged in canteen concentratedly. A solar hot water system is adopted for shower of the students and teachers who live in the school. A solar boiling water system may provide boiling water for teachers and students.

3 Ecological marsh and treatment of life sewerage

Utilizing the character of landform of the school life sewerage will be collected without dynamic to the southwest of the campus near ecological marsh and treated by ecological technology realizing the aim of no discharge.

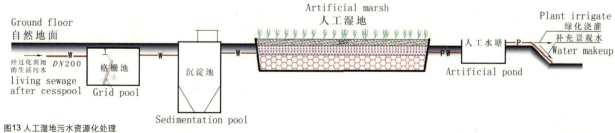

图13 人工湿地污水资源化处理
Figure 13 Sewerage recycling treatment via artificial marsh

主要技术经济指标一览表

序号	名称	单位	数量
1	总用地面积	hm²	2.74（含代征城市道路）
2	总建筑面积	m²	6551
2	教学及教学辅助用房建筑面积	m²	3712
2	宿舍建筑面积	m²	2284
2	食堂建筑面积	m²	555
3	道路广场面积	m²	3930
4	运动场用地面积	m²	5600
5	绿地面积	m²	10010
6	容积率		0.24
7	绿地率（不含运动场用地）	%	36.5
8	建筑密度	%	11.9
9	汽车泊位数	辆	10
10	自行车停车数	辆	100

Main technical & economic indicators

No.	Name	Unit	Quantity
1	Total area of the site	hm²	2.74 (including part of town road)
2	Total area of buildings	m²	6551
2	Floor area of teaching and office building	m²	3712
2	Floor area of hostel	m²	2284
2	Floor area of canteen	m²	555
3	Area of road and square	m²	3930
4	Area of playground	m²	5600
5	Area of greening	m²	10010
6	Floor area ratio		0.24
7	Greening rate(excluding playground)	%	36.5
8	Building density	%	11.9
9	Car parking spaces		10
10	Bike parking spaces		100

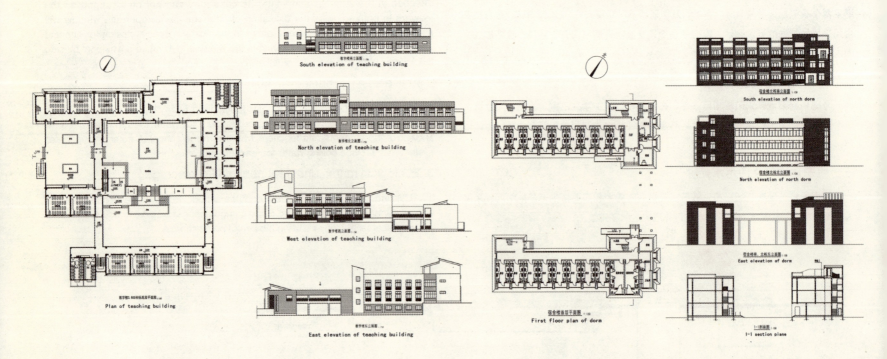

图14 教学楼方案图
Figure 14 Plan of teaching building

图15 宿舍方案图
Figure 15 Plan of dormitory

316

后记
Postscript

　　国家住宅与居住环境工程技术研究中心和中国可再生能源学会太阳能建筑专业委员会作为本次竞赛的承办单位，除了为灾区学校重建工作提供太阳能建筑竞赛活动的相关成果，还积极与各方联系，争取将获奖作品实地建设，为灾区的孩子们建设一座在太阳能技术支持下、充分利用清洁能源的"阳光小学"。这个想法得到竞赛活动冠名单位台达环境与教育基金会的大力支持，他们与国务院台湾事务办公室积极沟通，与台达电子集团、四川省绵阳市台湾事务办公室多次交流，决定将台达电子集团于汶川地震后捐赠的1000万元人民币，定向援助四川省绵阳市涪城区杨家镇小学，并建议按照竞赛活动获奖作品方案进行全校异地重建。

　　2008年12月4日，"阳光小学"建设捐赠单位台达电子集团中国区行政副总裁曾纪坚、见证单位四川省绵阳市台湾事务办公室副主任贺俊以及代表受助方的四川省绵阳市涪城区政府副区长蒋丽英三方在京共同签署了援建项目协议。

签约仪式

签约后集体留影 (Photo after subscription ceremony)

　　根据三方达成的重建计划，杨家镇"阳光小学"正在根据实际建设要求、场地条件和获奖作品851号，由中国建筑设计研究院修改完善设计。杨家镇阳光小学将于2009年6月开始付诸施工建设，并在2010年春季开学前建成并启用。

　　援建项目协议的顺利签订，标志着本届太阳能建筑设计竞赛首次实现了获奖作品的实地建设，使纸质作品成为可供观摩、可经受运行检验的建筑实体，不仅使国

　　As the operator expecting providing relevant achievement of solar building design competition for the reconstruction of the school in disaster area, China National Engineering Research Center for Human Settlement (CNERCHS) and Special Committee of Solar Buildings, Chinese Renewable Energy Society have also done their effort to contact relevant units actively for putting awarded works into implementation and constructing a "sunshine primary school" supported by solar energy used with clean energy for the children in disaster area. This idea gained the support of sponsor, Delta Environmental & Educational Foundation. It has contacted with Taiwan Affair Office of the State Council, made intercourse with Delta Electronics, Inc. and Taiwan Affair Office of Mianyang Government, Sichuan province and finally decided to put RMB 10 million contributed by the Group after the Wenchuan earthquake aiding the project of Yang Jia Zhen School of Fucheng District, Mianyang and suggested to reconstruct in another place according to the awarded scheme of the competition.

　　On December 4, 2008 Mr. Zeng Jijian, vice-administrative president of China District, Delta Electronics, Inc. as the donation side, Mr. He Jun, vice-director of Taiwan Affair Office of Mianyang as the witness and Ms. Jiang Liying vice-director of Fucheng District Government, Sichuan province as the aided side subscribed the agreement of the aid project of "Sunshine primary school".

　　Based on the reconstruction plan made by three sides, the design of Yang Jia Zhen "Sunshine primary school" is in revise and perfect by China Architecture Design & Research Group according to actual demands of the construction, site condition and the awarded works No. 851. The project will be started in construction in June, 2009 and finished and taken into use before term beginning in the spring, 2010.

　　The subscription of the agreement of the aid project symbolizes that this competition has realized putting awarded works into implementation for the first time. That transferring works in paper into building entities, which may be viewed, emulated and examined by operation, will push the demonstration platform of the achievement of the international solar building design competition to a new and high level. It will play an active role in aspect of solar application on school buildings.

　　One thing worthily being mentioned is that combining with the reconstruction after the earthquake, Yang Jia Zhen School conduct a match of composition and

际太阳能建筑设计竞赛成果的展示平台上了新高度，而且必将对太阳能在学校建筑上的应用起到积极的推动作用。

值得一提的是，杨家镇小学还结合学校灾后重建开展了《我心中的阳光小学》征文和绘画比赛，这项活动不但展现了孩子们丰富的想象、乐观的精神和烂漫的童真，而且也体现了阳光小学建设的宗旨，那就是让太阳能等可再生能源利用融入孩子们的生活方式中去，通过孩子们的亲身体验，把节约能源和保护环境的生态文明种子播种在他们的心灵深处，并传播到家庭等社会的每个角落。

drawings titled with "Sunshine primary school in my mind". It not only opens out abundant fancy, affirmative spirit and brilliant puerility of the children but embodies the tenet of constructing sunshine schools, which is to introject the idea of the application of renewable energies like solar energy into the life style of children and through the experience of the children themselves to seed ecological civilization concerning energy saving and environmental protection into their hearts deeply and even every corner of families and the whole society.

杨家镇小学的同学们所企盼的新学校

(New Yang Jia Zhen School being expected by students)